Charles Horetzky

Some startling facts relating to the Canadian Pacific Railway and the north-west lands

Charles Horetzky

Some startling facts relating to the Canadian Pacific Railway and the north-west lands

ISBN/EAN: 9783337131739

Printed in Europe, USA, Canada, Australia, Japan

Cover: Foto ©Andreas Hilbeck / pixelio.de

More available books at **www.hansebooks.com**

SOME STARTLING FACTS

—RELATING TO THE—

CANADIAN PACIFIC RAILWAY

—AND THE—

NORTH-WEST LANDS,

—ALSO—

A BRIEF DISCUSSION

REGARDING

THE ROUTE, THE WESTERN TERMINUS

—AND—

THE LANDS AVAILABLE FOR SETTLEMENT,

—BY—

C. HORETZKY.

Ottawa:
PRINTED AT THE OFFICE OF THE "FREE PRESS," ELGIN STREET.
1880.

Entered according to Act of Parliament of Canada, in the year 1880, by Horetzky, in the office of the Minister of Agriculture.

Prefatory Remarks.

Various indications point to the existence of a wide spread and rapidly growing feeling of dissatisfaction and distrust throughout the older provinces of the Dominion, with regard to the adopted route and construction of that portion of the Canadian Pacific Railway which traverses the "Rocky Mountain" and "Cascade" Zones, and terminates at Burrard Inlet.

The main difficulty in the construction of the Pacific Railway is encountered in the " Cascade " or Coast range, through which any line from the interior of the Continent must pass in order to reach the coast. All surveys made hitherto have been met by this grave obstacle. The "Yale," "Bute Inlet," "Kemsquit," and "Skeena" routes are all, owing to this impediment, well nigh impracticable, and the adoption of any one of them could only be justified, were it to lead to tangible advantages beyond. No such advantages exist, the entire seaboard being but the adamantine buttress of a mountain range, one hundred miles in width.

The seaboard again, although pierced by countless Inlets, and presenting on the map a most favourable appearance, offers in reality very grave obstacles to the mariner, because of the nearly universal dearth of good anchorages. On the whole mainland coast there is but one really good natural harbour—Port Simpson. All others have drawbacks in a more or less degree.

The writer claims to be able to point out a solution of the coast range difficulty, besides certain other advantages of paramount importance. The matter in the following pages is, to some extent, a mere index of facts culled from various railway reports, in order to elucidate the writer's arguments, in favour of a Northern route for the Pacific Railway through the "Pine River Pass" of the Rocky Mountains, preferably to that of the "Yellow Head Pass," as advanced in 1874 in a little work entitled "Canada on the Pacific."

The writer has been connected with the Pacific Railway surveys since the inception of the project, and during the past nine years has seen and examined much more of the North-Western country and of British Columbia, than perhaps any Engineer of Mr. Fleming's staff.

Besides having originated the northern route *via* the Pine River Pass, in opposition to the sneers of certain individuals identified with the Frazer River Line, the writer claims a special practical knowledge of the British Columbian coast, from the Alaskan boundary line, southwards, and has, therefore, no hesitation in giving his views to the public, he merely asks the reader to examine carefully the written testimony, and to trust to his own common sense for his deductions.

CHAS. HORETZKY,
Late of the C. P. R. Surveys.

OTTAWA, May 31st, 1880.

THE object of the present pamphlet is to place prominently before the thinking portion of the Canadian public certain facts bearing materially, not only upon the future prospects of the country at large, but also and by no means in a small degree, upon British Imperial interests, in so far at least as these interests may be vested in the Canadian Pacific Railway.

Acting upon the advice—presumably at least—of the Chief Engineer of the railway in question, the Government of Sir John A. Macdonald has taken the initiatory step towards carrying out the compact of 1871 with British Columbia. Construction has been commenced on the Yale-Kamloops Section—a length of 125 miles.

The Toronto *Mail* of the 6th May, in commenting upon the very serious question of routes, betrayed its misgivings by the utterance of the following apologetical remark:—

" It must be said that if a mistake should have been made in the " choice of the Burrard Inlet route, that mistake will have been made " without shame or blame to any one."

Exception may, perhaps, be taken to the above allegation. It is scarcely a matter of doubt that a mistake, and a very serious one, *has* been made. The article in question speaks of three seriously competing routes: the " Burrard," the " Bute " and the "Port Simpson " routes.

" After eight years of surveys pushed forward at great cost, and " with infinite labour to all concerned," the question has *not* been exhausted, and the testimony given in the following pages establishes beyond doubt that the true trans-continental route, and the true "Pacific" terminus of that route, have been most unaccountably lost sight of.

Six years ago, during an examination of the north-west coast of British Columbia, I discovered at the head of the "Kitimat" Inlet, or Douglas Channel, a small land-locked harbour, north from which stretched a beautiful valley leading directly to, and touching, the River Skeena at a point 75 miles above Port Essington. So much impressed were my little party and myself with the natural facilities of this locality towards the formation of a good harbour, and its adaptability for a terminus, that I made two attempts to discover access from this point to the eastern interior plateau. My efforts were fruitless, and after a five weeks' delay at the head of this Inlet, our sloop, the "Triumph," left for the Dean Canal.

All the above facts were duly set forth in my Official Report, dated from " Bellabella, North-West Coast B.C., November 15, 1874," but for reasons unknown to myself, that Report was mutilated most unmercifully, and the last portion of it, descriptive of the coast from Douglas Channel to Queen Charlotte Sound, entirely suppressed. Reference to pages 137, 144, of Mr. S. Fleming's Report of 1877, will afford the most convincing proofs of this statement. A foot note at bottom of page 144, states that the matter omitted is treated of in another appendix. I can only say that such is not the case. The only reference made elsewhere to the valley of the Kitimat—none is made to the harbour or head of the Inlet—is at page 111 of Mr. Fleming's Report for 1877, where Mr. Marcus Smith alludes to "*the wide and low valley stretching to the north affording an easy route to the Skeena River.*" His visit to that locality was made a few days previous to mine, but he did not go far enough up the Inlet to see the bay at the northern extremity.* Now, this place being but 140 nautical miles from Port Simpson, or, say ten hours steaming through the safest and finest channels in the world, *i.e.*, the "Grenville" and "Douglas" Channels, the former of which has been used for years, and safely navigated during the darkest and most stormy nights of winter by both British and American steamships bound up and down between Sitka and southern ports, being also accessible from the ocean by "Nepean" and "Wright" Sounds, having also the advantage of havens of refuge at two places in the Grenville Canal, viz: Klewnugget Inlet and Lowe Inlet, besides Kitkatlah Bay on the west side of the Douglas Channel, and "Coghlin Anchorage" under the lee of Promise Island, not to speak of Clio Bay situated five or six miles south from the mouth of the river Kitimat, which not only affords anchorage, but also facilities for the construction of a dry dock, I dare affirm that it is extremely well situated for a terminus, being as near to Yokahama, Japan, as Port Simpson, viz: 4,000 geographical miles, and 400 miles shorter than the southern passage from Burrard Inlet.

I shall refer further on, to the marine engineering works required to transform the upper end of the Douglas Arm into a good harbour, and shall now proceed to describe the features of the railway route from

---

*NOTE.—Mr. Marcus Smith was the officer in charge of the western section until 1377. He has an intimate knowledge of British Columbia.

Kitimat, eastward to a point on the now located southern line, known as "Livingston," distant 682 miles from Thunder Bay, on Lake Superior.

A circumstance which, more perhaps than anything else, commends this route to our consideration, is the extraordinary fact that the formidable coast range of mountains which necessitates such frightful expenditure on the "Yale Kamloops," "Homathco," "Kemsquit," and Port Simpson routes, can be passed upon this one, with nearly as little trouble, danger or cost, as upon an average prairie section, the valley of the Kitimat being several miles in width, of a nearly level character, and clothed with a magnificent forest of heavy spruce, hemlock, cedar, and other trees, amongst which crab-apple and maple may occasionally be seen.

From tide water at the head of Douglas Arm, the valley rises almost imperceptibly for about 20 or 25 miles, at the rate of four-tenths per hundred to the "Divide," near Lake "Killoosah" or "Lakelse," thence the descent to the Skeena, or to some point upon that river nearly opposite the "Kitsumkallum" River, is very gradual, the entire distance probably not exceeding forty (40) miles. The mouth of the "Kitsumkallum" River is, by Mr. Keefer's estimate, about seventy-five miles above Port Essington, or rather more than 100 miles from Port Simpson, within which distance the work of railway construction along the Skeena would be extremely heavy, the line proposed being carried in the river bed in many places where the mountain bases afford little chance for a road-bed. In more than a dozen places the precipitous slopes are swept by avalanches of the most dangerous character. The shores of Wark Inlet are but little better. The contrast between those routes is very striking, and, as has been shown, the distance between Kitsumkallum and the sea is very much more than doubled on the difficult and expensive Skeena line. Roughly estimating, for the sake of comparison, the cost of construction on that portion of the Skeena route at six million dollars, and that of the Kitimat at one-and-a-half million dollars, we have a difference in the first section from tide water of four-and-a-half millions dollars, not to speak of extra cost for maintenance and repairs, which would, of course, be very much greater on the Port Simpson, Skeena route. One-fourth of that difference would go far towards the formation of an excellent terminal harbour at the head of the Douglas Arm.

From a point opposite Kitsumkallum River, on the left bank of the Skeena, the distance, to Hazelton, or the "Forks," is little more than seventy

(70) miles. In this distance, according to the careful estimate of Mr. H. MacLeod, the grades would be very moderate, and the proportion of heavy work small, say about one-seventh, or ten miles, the balance moderate and light. *Vide* page 59, Mr. S. Fleming's Report of 1880.

At the 110th mile the line would attain, with very moderate grades throughout, an elevation above sea of nearly one thousand feet, upon the high terraces at the base of the Rocher Déboulé Mountain. Here it leaves the Skeena to follow the valley of the "Wotsonqua" for 117 miles, to the summit between the waters of the latter and those of the "Intaquah." Mr. H. J. Cambie, thus describes this section, at page 39, Appendix C. of Mr. S. Fleming's Report for 1878 :—

" The River Wotsonqua, from its mouth at the ' Forks,' up to the " Indian Village of 'Awkelget,' 27 miles, runs through a deep Canôn. " The works would be generally heavy, but some exceedingly so, with " stiff gradients and sharp curves. From 'Awkelget' upwards, the " valley is favorable for railway construction, works would be moderate, " with easy grades for about 90 miles, to the summit between the waters " of the Wotsonqua and Intaquah, distant 300 miles from Port Simpson, " and about 2,400 feet above sea level. Thence to the valley of the " 'Nechaco,' the works would be moderate, with easy grades."

One would infer from the foregoing description that in the 27 miles between the " Forks " and Awkelget, the line would follow the Canôn of the Wotsonqua. This would not be the case, the line would take higher ground upon terraces several hundred feet above the river level, and no great difficulty would be encountered in a large proportion of the 27 miles below Awkelget. Moreover, between the village of " Kitsigeuchlé "— situated 15 miles below the Forks of Skeena—and the latter, there is fairly level ground along the mountain bases, where the works would be quite moderate, and upon which a line can be carried with easy gradients up to a level entirely *above* the rough Canôn of the Wotsonqua. I have been over the ground, above and below the " Forks," and know this to be the case, so we must not assume the whole distance Mr. Cambie refers to —27 miles—nor anything like it, to involve heavy work. However, accepting his estimate, the balance of the distance to the Wotsonqua Summit, consists of moderate work with light gradients, for 90 miles.

From the Wotsonqua Summit, at the 227th mile from Kitimat, there is no difficulty as to works or grades, as far as the Stewart River (distance 95 miles.) *Vide* page 53, Mr. S. Fleming's Report for 1880.

From Stewart River to MacLeod Lake (distance, 75 miles) there will be about twenty (20) miles of heavy work, with grades in some places exceeding one per hundred, the balance will be moderate to light. (*Vide* pages 53 and 61, Mr. S. Fleming's Report, 1880.)

From MacLeod Lake to "Pine Pass" Summit, the distance is forty (40) miles, of which about one-fifth involves heavy work, in addition to some deep cutting at the approaches to the River "Parsnip," the balance will be of medium character, and grades not in excess of one per hundred. (*Vide* pages 52 and 53, Railway Report, 1880.) The "Pine Pass Summit is at the 437th mile from "Kitimat."

Mr. Cambie travelled from Hudson's Hope to Pine River, which he came upon 35 miles east from the summit. Of the portion he saw, seven miles are described as heavy. He also refers to the necessity for some protective works at a few precipitous points upon the river, but apart from these, he saw no serious obstructions, and anticipates none as far as the "Lower Forks," 75 miles east from Pine Pass Summit.

Mr. MacLeod describes the next section, *i. e.*, from the "Lower Forks" of Pine River, eastward to Smoky River, a distance of 139 miles. He estimates that in this portion there may be, perhaps, 20 miles of heavy work, the balance light and moderate. (*Vide* pages 64 and 65, Mr. S. Fleming's Report, 1880.)

Between Smoky River and Lesser Slave Lake, no member of the Peace River Expedition took the direct line, Mr. H. Cambie having strayed to the south-east, while Mr. MacLeod diverged purposely towards the Athabasca River. The former, however, travelled over the trail between Slave Lake and Peace River, a distance, he estimates to be 55 miles. Mr. Cambie anticipates however no difficulties whatever, between the points in question. The estimated distance is 60 miles.

From the western end of Lesser Slave Lake, the line would follow the south shore to its outlet, descend Slave River, and cross the Athabasca, on a direct line for Lac La Biche, or some point slightly south of it. Mr. Cambie saw nothing of this section. I did however, in 1872, and from Mr. Gordon's cursory observations regarding the low flat land adjacent to Little Slave River, and his account of the country from the Athabasca Landing eastward, it is certain that the works throughout will be of very moderate character.

Between the Athabasca and the meridian of Lac La Biche, there are, according to Mr. Gordon, some 21 miles of rather poor soil, but the country becomes very rich on going east. The estimated distance from Lesser Slave Lake's western extremity to Lac La Biche is about 200 miles, nearly all remarkably favorable for railway construction. A crossing of the Athabasca can be effected anywhere, the waterway would not exceed 600 feet. A very large proportion of the land in this section is fit for agriculture and the growth of wheat.

Between Lac La Biche and Livingston, the distance, with a handsome allowance for deviations, is 470 miles, over a gently undulating country, of which by far the greater portion is better adapted for agricultural purposes than any on the Saskatchewan. In this connection, the reports of Messrs. M. Smith, O'Keefe, Eberts, Macoun and King, in regard to its agricultural capabilities, may be of interest.

The first named gentleman, in speaking of this large section, says, at page 47, Report of 1878 :—" Following up the valley of the Swan River " about 80 miles, the line would take a direct course for the Saskatchewan, " near Fort à la Corne." " The land in the Valley of Swan River " is reported by the Surveyors to be very rich and of considerable ex-" tent ; the soil on the Basquia Hills is also reported good ; while the " belt between these hills and the Saskatchewan, extending from the " Prince Albert settlement, above the Grand Forks, down to the Old " Fort, a distance of over 90 miles, is exceedingly rich land."

" From the Saskatchewan, the line would be nearly direct to the " foot of the Lesser Slave Lake, skirting the north side of the Moose " Hills, on the water shed of the Beaver River and passing the south " end of Lac La Biche or Red Deer Lake. Low ranges of hills skirt the " north bank of the Saskatchewan from a point a few miles above Fort " Carleton nearly to Victoria ; these are partially covered with groves of " aspen and willow ; the soil is generally light, but is well supplied with " streams of clear water ; the pasturage is good, especially in the " neighbourhood of Fort Pitt."

" Between these hills and the River the soil is generally sandy, " and there are numerous salt or alkaline lakes ; but immediately north " of the hills, the soil is stated by the officers of the Hudson's Bay " Company, to be very good. There are numerous fresh water lakes, " abounding in white fish."

" The writer drove out 16 miles north-west of Carleton, and found " the character of the country gradually improving, as he had been led "to expect from the description of it given by Mr. Clarke, the Chief

"Factor at the Fort, who has spent many years in this district. An Excursion was also made from Fort Pitt to Lac La Biche. The south slope of the Moose Hills, where the trail runs, is covered with a dense grove of aspen; but in crossing the west end of these hills, a magnificent prospect opened out. Stretching away to the east, north and west, as far as the eye could reach, there appeared a vast, undulating, grassy plain, rising in places into softly rounded hills, dotted and intersected with groves and belts of aspen mixed with spruce and tamarac and clumps of willows. This appears to have been formerly forest, which has probably been destroyed by fire, decayed trunks of large trees being found on the hill sides. In the hollows, however, there is sufficient timber left for railway and domestic purposes. The altitude, taken at several points, averages about 1,700 feet above the sea level."

"During three days, whenever the trail was left, great difficulty was found in forcing a way through thick masses of grass and pea-vine, three to four feet in height, and sometimes reaching nearly to the horses' backs. As Lac La Biche was neared, the country became more wooded, and the track lay through long glades between belts of poplar and willows, passing a number of small fresh water lakes."

"There is a Roman Catholic Mission at Lac la Biche, where they produce excellent wheat, barley, oats and all kinds of vegetables; there are about 40 families settled round the Lake, chiefly half-breeds, engaged in the fur trade, and only cultivating enough of cereals and vegetables for their own use. Between this point and the Lesser Slave Lake, the Line crosses the River Athabasca. This country has not been explored for the Railway."

"The line would follow either the south or north shore of Lesser Slave Lake, as might be determined by the Surveys. After passing that lake, it enters on a vast region of great fertility, extending far northward on both sides of the Peace River, and westward to Pine River, which falls into the Peace near Fort St. John."

"By this route, what is termed the fertile belt, or wheat-producing country, extends nearly three hundred miles farther to the west before the Rocky Mountains are reached than by the route over the Yellowhead Pass; a corresponding reduction being made in the breadth of sterile country to be crossed in the Rocky Mountain district."

The total length of line just described, from Kitimat to Livingston, is 1,881 miles.

From English Bay, Burrard Inlet, to the same point (Livingston), the measured distance is 1,281 miles. Thus, there is a difference of 100 miles in favour of the southern route.*

---

* NOTE.—The reader should read carefully the reports of Messrs. McLeod and Cambie. Railway Report of 1880.

A line from Kitimat, via Hazelton, Babine, the Kotsine Pass, and the rivers Omenica and Peace, would only be some 30 miles longer than the Burrard Inlet line. The "Pine Pass" route is, notwithstanding its greater length, preferable. Although being 70 miles longer than the line via the "Peace River Pass," its construction would be less costly, the gradients throughout much more moderate, and it would open up a greater extent of cultivable and pastoral land. We shall, therefore, confine the discussion to the "Pine Pass" route, and for a description of the "Kotsine Pass" and Peace River Line, the reader can, if desirous, find one at page 75, Appendix No. 5, of Mr. Fleming's Report for 1880, where an account of the writer's examination, during the season of 1879, is given, *in extenso*.

At page 8 of the last Railway Report, Mr. Fleming institutes a comparison of the Port Simpson, Peace River; Port Simpson, Pine River; and Port Simpson, Yellow Head routes. They are represented there as being, respectively, 190 miles, 255 miles, and 225 miles, longer than the Burrard Line. With all due deference to that gentleman, I am obliged to differ from him in this matter. The comparison is unfair, Fort Saskatchewan, the point to which those lines are referred, not being properly common to all, and certainly not being on the "Pine" or "Peace" River lines.

Taking Livingston as the common point for all, the real difference between Port Simpson lines, via Peace and Pine Rivers, and the Burrard route, is, respectively, 100 miles and 170 miles, while from the "Kitimat," the actual difference is further reduced to 30 miles and 100 miles, in the respective cases, as before shown.

Giving to the southern, Burrard Line, then, its undoubted advantage in distance (100 miles) ovea the Pine River route, we shall institute a rough comparison of the engineering features pertaining to each of the two lines now in discussion.

Between Livingston and the Yellow Head Pass, a distance of 771 miles, there are 75 or 80 bridge structures over streams and dry ravines, some of which are of very great magnitude. Notably, two crossings of the Saskatchewan, and those of Eagle Creek, the Pembina, MacLeod, Athabasca and Assiniboine Rivers, the width of waterways varying from 40 feet to 1600 feet. The approaches to those are also very difficult, in some cases involving excessive excavations.

There are also 2425 feet of tunnelling near the Yellow Head Pass. Between the Yellow Head Pass and Fort Moody (492 miles), there is an aggregate of 2¼ miles of tunnelling, and there are also 174 bridges with spans varying from 40 to 275 feet. The gradients are, however, kept within the maximum of one per hundred. Between Yellow Head and the village of Yale (403 miles), fully 70 per cent. of the distance will entail very heavy work, and in the portion now under contract (125 miles), the work throughout is excessively heavy, and will cost at least $100,000 per mile, equal, for that insignificant distance, to $12,000,000. For half the distance between Yale and Port Moody, the work is classed as heavy. In the whole distance between the Yellow Head and Port Moody (492 miles), there will be more than 320 miles of heavy work. The cost of this British Columbian section, from Rocky Mountain Summit (Yellow Head Pass), to the sea, is estimated, or rather underestimated, at thirty-two million dollars.

Taking up the latest Reports of Messrs. Cambie and MacLeod, we gather that upon the "Kitimat, Pine Pass" route, from a point on the Skeena, opposite the Kitsumkallum River, *via* Hazelton, the Wotsonqua Valley, and the Pine River Pass, to the last named summit of the Rocky Mountains, those gentlemen estimate that there are about seventy-five (75) miles of heavy work.

The corresponding section upon the southern line, as already shown, gives 320 miles of heavy work.

Between the Pine River Pass and the Smoky River, they report 30 miles of heavy work, including three miles of heavy excavation on each side of the last named stream. From Smoky River, eastward, to the meridian of Lac La Biche, a distance of 260 miles, the country is generally so favorable, that the proportion of heavy work upon that section is but trifling. The streams to be crossed upon this route, are fewer than upon the southern line, and far less difficult to bridge, the principal among them being the Athabasca (600 to 700 feet), the Smoky River (750 feet), the accent Échaland (800 feet), the Mud (400 feet), the Pine (500 feet), the Parsnip (600 feet), the Stewart, (600 feet). Between Lac La Biche and Livingston, (470 miles,) the country is, according to the best authority, quite as favorable, probably more so, than the corresponding portion west from Livingston, upon the southern line.

Taking the mileage upon the respective routes, from Rocky Mountain Summit to Pacific tide water, and calculating the heavy work given in the reports at $75,000 per mile, we find that upon the Burrard Inlet line there are :—

320 Miles of Heavy Work	@ $75,000....	$24,000,000
173 " Moderate "	50,000....	8,650,000
493		$32,650,000

And upon the "Kitimat Pine Pass" line,

75 Miles of Heavy Work	@ $75,000....	$5,625,000
362 " Moderate "	50,000....	18,100,000
437		$23,725,000

Giving a difference in cost of $8,925,000, in favour of the northern Pine Pass Line. But on the latter, there are, at the lowest calculation, 20 miles in the valley of the Kitimat, and 90 miles in that of the Wotsonqua, 110 miles in all, far easier to construct than the lightest portion of the Frazer River line, so that we may safely assume the total cost of the northern, British Columbian section (437 miles) at twenty-two millions, or say a difference of ten millions dollars, between the two lines. The reader can here turn to page 57, Appendix E, Railway Report of 1878, for Mr. Cambie's estimates of cost.

It is not so easy to form comparisons of cost from Rocky Mountain Summit, *eastward* to the common point, Livingston, as no systematic examinations have yet been made east from Lesser Slave Lake. It is, however, certain that the northern line will be found quite as easy of construction, probably far more so, than the southern, owing to the fact that *east* from the Smoky River to Livingston, the general profile of the country is more uniform than on the southern line, and that there are fewer streams to cross, and *only one crossing of the Saskatchewan* necessary.

At page 48, Report of 1878, Mr. Marcus Smith says of the engineering features of the Northern line, via Pine Pass :—

"It is difficult to form even an approximate estimate of the cost of
" construction without surveys, but the explorations across the Rocky
" Mountains show that a very great reduction can be made on the rock
" and earth excavations by the line through Pine River Pass as compared
" with the line by the Yellow Head Pass. On the Summit there will

" be about eight miles of heavy work. On the West side of the pass to
" the point of junction of the two lines the works will be very light, *and*
" *the cost probably not more than half that on the other line, mile for mile.*

" The bridging on both lines will be rather heavy in the centralor
" prairie region and on the eastern slope of the Rocky Mountains, but
" the number of very large structures will be much greater on the
" southern than on the northern route.

" *On the whole, the cost of the works of construction on this route may*
" *be safely estimated, so far as our examination extends, as very considerably*
" *below that on the other route.*"

An examination of the latest reports shows that, with the exception
of a few miles on the section between "Stewart" and "MacLeod" Lakes,
and at the approaches to the Smoky River, the grades can be kept within
a maximum of one per hundred, while the features of the country
passed by the northern line are so favorable, compared with the
southern route, that curvature will probably be far less, at least within
the Rocky Mountain and Cascade Zone, *i. e.*, upon the Pacific Slope.

From tide water, at the head of Douglas Arm, to Livingston, the
general profile of the northern line is the most remarkable upon the
North American continent. In the 625 miles west from the Smoky
River, there are only four summits, ranging, in altitude, from 2,400
feet to 2,750 feet above sea level. The "Pine Pass" summit is placed
at an elevation of 2,800 feet, an estimate which I have reason to think
slightly in excess of the reality.

Having now compared to some extent the engineering features of
the two lines, the capabilities for settlement—a matter of exceeding importance—of the regions traversed by the rival routes, demand investigation. On the southern line, from Livingston to the River Pembina—some
fifty miles west from Edmonton—which may be set down as the western
limit of the good, agricultural land on that parallel, there is a fair
proportion of land fit for settlement and pastoral purposes.

On the northern route, between Livingston and the meridian of the
Pembina River, the whole country, with the exception of a small portion
near the Athabasca, is well adapted for settlement. [*Vide* reports of
Messrs. Marcus Smith, O'Keefe, Eberts, Macoun King and others.]
At Lac La Biche, wheat has been for many years an unfailing crop;
tobacco has also been very successfully raised.

Indeed, an inspection of the reports of the above named gentlemen, and our previous knowledge of that region, leads to the inevitable conclusion that the northern line, as proposed in 1874, passes through a country, equal for purposes of settlement, to that south of the North Saskatchewan. The proportion of good land is as great, while wood for fuel and other purposes is more abundant, fresh water lakes of more frequent occurrence, the aspect of the country more prepossessing, and its altitude less, than on the southern line.

Of the country upon the southern line, between Lake St. Ann (a few miles east from the Pembina,) and the Rocky Mountains, Mr. H. MacLeod, C.E., speaks in the following strain. [See page 202, Report 1877]:—

" The soil is principally heavy clay, and in places, sand. Frequent " rains, not being permitted to sink into the ground, by the compactness " of the clay, form large areas of muskeg. There are a few places where " the soil in the valleys is fair."

Ice has been seen in those muskegs referred to by Mr. MacLeod, during the month of August. The Reverend George Grant, author of "Ocean to Ocean," remarks of this region, at page 193 :—" Poor, scrubby timber, the land cold and hungry." Idem at page 200: "Brush has decided autumnal tints," August 31, '72. " Country of a decided poverty-stricken look." " Miles of muskeg."

The reverend gentleman was decidedly right. The whole region between the Pembina River and Jasper House is cold and clayey, and covered with muskeg for long stretches. Its altitude varies from 2,500 feet to 3,400 feet above sea. It is, as expressed years ago in "Canada on the Pacific," "cold, swampy, and unfitted for settlement."

The distance from Lake St. Ann, the western limit of the good land or fertile belt upon the southern line, to Jasper House, is about 160 miles.

From Jasper House to a point near Kamloops—nearly 300 miles —the country is totally unfit for settlement. I have never passed over this section, but shall quote from authentic sources. Mr. George Keefer, one of Mr. Fleming's engineers, says at page 355, Pacific Railway Report of 1877:—In the Fraser River Valley, "but of the 12 months, two and " even less would be all that could be relied upon as exempt from frosts, " but few cereals could be raised in this locality. The amount of arable " land in the valley of the Frazer is so small that it is hardly ever likely " to be taken into consideration."

Mr. Marcus Smith says at page 45, Report of 1878:—

" From the Pembina River across the Rocky Mountains' to a point
" near Kamloops—420 miles—is totally unfit for settlement. There is
" another length of 100 miles in the canyons of the Thompson and
" Frazer in a similar condition. So that from the River Pembina, on the
" east side of the Rocky Mountains, to the proposed terminus at Port
" Moody, a distance of 679 miles, there are 520 miles on which there is
" no land fit for settlement, and on the balance most of the land of any
" value is taken up ; in all this distance, therefore, there will scarcely be
" an acre within 60 to 100 miles of the line at the disposal of the Govern-
" ment for Railway purposes. The works, moreover, will be generally
" heavy and costly."

The Reverend George Grant says at pages 292 and 321 of " Ocean
to Ocean : "—

" Were we to judge from what we have seen of the country along
" the Frazer and Thompson Rivers, the conclusion would be forced on
" us that British Columbia can never be an agricultural country."
" But the greater part of the mainland is a *Sea of Mountains*, and
" the province will have to depend on its other resources for any large
" increase of population." (*Idem* at page 321.) " The little that we
" saw of the mainland of British Columbia, does not warrant us to say
" much about it as a field for emigrants."

The discussion of the agricultural capabilities of British Columbia,
will be resumed further on. We shall at present return to the Eastern
side of the Rocky Mountains. Between Lac la Biche and the western
extremity of Lesser Slave Lake, there may be, perhaps, 25% of poor
soil, but from the latter, westward to the " Middle Forks" of Pine
River, a distance of 225 miles, the northern line will pass through a
highly favoured portion of the North-West. Some years ago, when
I mooted the project of the Pine River Pass route, I was, to a certain
extent, aware of the great dissimilarity between the lands on each
route. Now, I am able to quote from other reliable sources, from entirely disinterested persons, and from Mr. Macoun, a strong partisan
of the southern line, who, despite his predilection cannot speak otherwise
than in the most extravagant terms of the Peace River region.

At page 113 of Mr. S. Fleming's Report for 1880, Doctor Dawson
thus defines the boundaries of the southern portion of the Peace River
country :—" With the ' Middle Forks' of the Pine River (only fifty miles
" east from the summit of the Rocky Mountains, in latitude $55\frac{1}{2}°$ N.), as a
" western limit, the region now to be described may be considered as
" bounded to the north by the 57th parallel to its intersection eastward

" with the Peace River. Thence the boundary may be assumed to
" follow the Peace River southward to the mouth of 'Heart Brook,' near
" the confluence of the Smoky River. Thence to run south-eastward to
" the extremity of Lesser Slave Lake, to follow the western border of the
" hilly region lying to the south of the lake, to the Athabasca River;
" thence to follow the Athabasca westward to the foot-hills, and skirting
" the foot-hills to run north-westward to the first mentioned point on
" Pine River. The tract included in the above limits has an area of
" about 31,550 square miles, and by far the greater portion of this area
" may be classed as fertile."

The above definition of boundaries agrees, so far as it goes, with those given by myself at page 228, "Canada on the Pacific." Doctor Dawson goes on to say:—" The soil of this district may be described
" as a fine silt, and not dissimilar from the loess-like material con-
" stituting the sub-soil of the Red River Valley in Manitoba. As evi-
" denced by its natural vegetation, its fertility is great. The total area
" of prairie land west of the Smoky River may be about 3,000 square
" miles. The total area of land with soil suitable for agriculture, may
" be estimated as, at least, 23,500 square miles. The luxuriance of the
" natural vegetation in these Peace River prairies is truly wonderful,
" and indicates not only fertility, but also sufficient rainfall. It
" may be stated at once that the ascertained facts leave no doubt
" on the subject of the sufficient length and warmth of the season to
" ripen wheat, oats and barley, the only point which may admit of ques-
" tion being: to what extent the occurrence of late and early frosts may
" interfere with growth."

Professor Selwyn says, at page 62, Geological Report of 1875:—

" I consider it a region (the Peace River country above Fort St.
" John,) far fitter for settlement than much of the Saskatchewan
" country." We are now in the midst of September (1875). The ther-
" mometer has only once reached 32° Fahrenheit. As a contrast to this, it
" will be seen in my Report upon the Saskatchewan country in 1873, that
" in the region about Edmonton and Victoria, two degrees further south,
" and about the same elevation, the thermometer fell, on the 4th Sep-
" tember, to 28°; on the 6th, to 24°; on the 11th, to 20°, and again to
" 20° on the night of the 23rd. At page 51, Professor Selwyn says:—
" We came upon a fine level country covered with the richest herbage,
" of astonishing luxuriance. I have seen nothing in the Saskatchewan
" country that at all equals it, both soil and climate are better here.
At page 68, he remarks:—" I have no hesitation in saying that,
" *the Pine River Route is probably in every respect the best in the interest of*
" *the Railway and of the country at large.* It will be found that by 'Pine
" River Pass,' the line could be carried almost the whole distance,
" through a magnificent, agricultural and pastoral country.

Mr. John Macoun, botanist, says at page 154 of the Geological

Report of 1875. He is speaking of the country near St. John's on the Peace River. "*25th July, 1875.* The oats stood five feet high. For
" nine miles the distance travelled, the country was covered with the
" most luxuriant vegetation. It would be folly to attempt to depict the
" appearance of the country, it was so much beyond what I ever saw
" before. The soil must be exceedingly rich to support such a growth
" year after year, and the early summer temperature must be high,
" for the vegetation to be so far advanced at this period. All the cul-
" tivation at St. John is on the terrace, immediately above the spring
" flood level, *but there is no reason why cereals should fail on the
" plateau above as the soil is there, if anything, better.* I never wit-
" nessed such an astonishing growth of herbaceous plants. The *flora*
" of this region is almost identical with that of Ontario. The winter
" is actually shorter on Peace River than in Manitoba, and the record
" shows, that, twelve hundred miles north-west from Fort Garry, a
" milder temperature prevails in autumn than at that point."

In his Report of 1874, Mr. Macoun, also, remarks :—" I am satis
" fied that wheat will succeed here (Lesser Slave Lake), as I think there
" is a higher summer temperature here than at Edmonton. What I saw
" of the Peace River country (at least a distance of 200 miles), was
" the best land I had seen anywhere. Here is a strip of country, over
" 600 miles in length, and at least 100 miles in breadth, containing an
" area of 60,000 square miles, which has a climate in no way inferior to
" that of Edmonton."

" Regarding the quality of the soil throughout the entire region,
" my note-book is unvarying in its testimony. It was principally clay
" loam, five feet in depth where exposed, but owing to the clay sub-soil,
" it is practically inexhaustible."

On referring to the reports of Messrs. H. J. Cambie and D. M. Gordon, we find that, despite preconceived ideas and strongly prejudiced though they were in favour of the route *via* the Yellow Head Pass, those gentlemen are perforce unable to speak but in the highest terms of the Peace River country. One little bug-bear they did find, and they endeavoured to make the most of it. The country would be perfect but for the nocturnal summer frosts which are experienced occasionally. They found wheat growing well in the bottom of the Peace River Valley, 600 feet below the general altitude of the country, and, because they saw no crops on the plateau (no one has yet settled there), they fancy that wheat culture there will be a failure, the cause being, as they say, a colder temperature upon the plateau than in the valley below. Had they taken simultaneous observations for temperature on plateau and in valley, they would probably have found on clear, calm nights, preceded by strong westerly winds (the occasions when those frosts usually occur), the air actually colder in the bottom of the deep valley than above, the heavier, and consequently, colder air being beneath. Every farmer knows this to be the case, low bottoms being generally more subject to frost than higher land. The difference of altitude in this case is not sufficiently great to bring into operation the general law, " that air temperature decreases in inverse ratio to the increase of " altitude. In fact, frosts rarely injure vegetation, unless both air and soil are saturate1 with moisture, and the Peace River country is proverbially dry. It is known that, in the State of Minnesota, the mercury sometimes falls to 20° Fahrenheit without doing injury.

Speaking of Hudson's Hope, Mr. Cambie says of the May frosts:—
" It was said by the people there, that the frost was confined to the
" valley, and did not extend to the plateau."

Mr. Gordon, who never left the beaten trail, remarks at page 103, Railway Report, 1880:—

" It might, therefore, be premature to pronounce even the most
" fertile portions of this plateau suitable for the growth of grain. Yet,
" various considerations seem to warrant the conclusion that climatic
" conditions are not less favourable on the plateau than in the valley.
" Frost sometimes occurs in the valley, when not felt on the plateau."

Doctor Dawson, an unprejudiced person, and a better authority, remarks at page 117, Railway Report, 1880 :—" In my diary of September 5th, I find the following entry : "—" Aspens and berry bushes about
" the Peace River Valley now looking quite autumnal ; on the plateau,
" 800 ft. higher, not nearly so much so. This difference appears to be
" actual." In October, 1872, Mr. Horetzky remarked the same circum-
" stance."

At page 9 of his last Report, Mr. S. Fleming remarks, with reference to the Peace River country :—" *The explorations do not establish*
" *beyond question its adaptability for the systematic growth of the higher*
" *cereals.*" Again, at page 10 :—"No frost was experienced at Ed-
" monton in last August, a fact which suggests that the Peace River
" District *cannot be considered equal to the Saskatchewan in point of*
" *climate.*"

At page 119 of the same Report, Doctor Dawson, in speaking of the country around Edmonton on the Saskatchewan, says :—" *The climate,*
" in its great diurnal and annual range, *corresponds exactly with*
" *that of the Peace River country.* Fort Saskatchewan is situated
" on the brow of the Saskatchewan Valley, about seventy feet above
" the river, and therefore, probably less liable to frosts than either
" the bottom of the river valley, or extensive flat tracts of plain where
" there is little circulation of air. *It would be premature to allow that*

" *the climate of the Peace River is inferior to that of the region about
" Edmonton on the Saskatchewan.*"

While on the subject of frosts, let us again refer to "Ocean to Ocean." At page 178, the author says, he is speaking of the region around Edmonton, on the Saskatchewan : " The remaining difficulty
" is the recurrence of summer frosts. These are dreaded more
" than anything else. At one place, in June or July ; at another
" in August, sharp frosts have nipped the grain. At Edmonton there
" is invariably a night or two of frost, between the 10th and 20th of
" August. At Victoria and Fort Pitt, to the east, at St. Albert
" and Lake St. Ann, on the west, the grain has suffered more or less,
" frequently, from the same cause. This enemy is a serious one, for,
" against it man seems powerless. But, admitting to the full that
" there are such frosts, that no improvement will ensue on the general
" cultivation of the land, the draining of bogs, and the peopling of the
" country, other crops than wheat can be raised. It is only fair to
" the country to add that the power of these frosts to injure must be
" judged, not by the thermometer, but by actual experience. It is a
" remarkable fact that frost which would nip grain in other countries
" is innocuous on the Red River and on the Saskatchewan. Whatever
" the reason, the fact is undoubted."*

To-day, despite the frosts, wheat succeeds to admiration at and around Edmonton, and elsewhere on the Saskatchewan, thus proving the groundlessness of the fears expressed by Mr. Fleming's Secretary in the the foregoing quotations. Why, then, not apply the same arguments in the case of the Peace River country ? Why lay such stress on the occurrence of frosts as experienced by Messrs. Cambie and Gordon ? What applies in one case will, also, in the other. If the frosts do not injure wheat crops in one part of the North-West, they will not in another, possessed of a similar climate, and Doctor Dawson tells us that *the climate of the Peace River region corresponds exactly with that of the Saskatchewan.* Professor Selwyn is of the opinion that the Peace River is *even better adapted* for agriculture than the Saskatchewan ; while Professor Macoun takes similar ground, and they all admit the superiority of the Peace River region in other respects. Why, wheat succeeds admirably at Fort Simpson and Fort Laird, in latitude 62°, four hundred miles further north than the portion of the Peace River country seen by Messrs. Cambie and Gordon, and, if I mistake not, samples of wheat grown near Lake Athabasca took a prize at the Philadelphia Exhibition. It is useless to pursue the argument further, the trivial objections raised by Messrs. Cambie and Gordon, must fall to the ground be

---

*NOTE.— The reason is easily explained. The frost is innocuous owing to the extreme dryness of both soil and atmosphere.

neath the weight of testimony quoted above, and when they advance the ridiculous argument of a warmer climate in the bottom of the Peace River Valley, than on the plateau above, one is inclined to doubt, if the selection of those gentlemen to investigate a subject of such grave import to the country at large, and to the taxpayers in particular were altogether, wise and proper.

Sufficient evidence has now been adduced to convince even the most sceptical that a mistake in the selection of the trans-continental route has been made. It may be added, that the northern line will open up an immense system of navigation, from its crossing point upon the Athabasca River, down to the " Mackenzie," the great lakes, the Arctic Fisheries, and to the immense mineral regions of the Lower Athabasca and other streams.

Doctor Dawson tells us at page 112 of the last Railway Report, 1880, that *the greatest connected region susceptible of cultivation in British Columbia*, is in the flat country of the lower " Nechaco " basin. This region lies immediately south from, and adjacent to, the Pine Pass line. Mr. Macoun speak highly of this region, also, and I shall here quote what he says of its soil and climate :—(*Vide* pages 134, 135, Geological Report of 1875.)

" Looking back over the 146 miles which lie between Fort St. James
" and Quesnel, I am struck with the resemblance of the flora to that of
" the forest region west of Lake Superior. There is not a species in the
" whole distance which in any way indicates either an alpine or a boreal
" climate, except *Vaccinium myrtilloides* and *Empetrum nigrum*, and
" these were only observed once. The valley of the Nechaco has an
" exceedingly rich soil on both sides where the trail crosses, and possibly
" this extends for many miles both above and below. The valley of
" Steward's River is not wide where we crossed it, but it is very rich,
" and there is no doubt whatever, in my mind, but that after the two
" rivers unite, the valley all the way to Fort George is rich and fertile,
" and well suited for settlement. From the crossing of Stewart's River
" to Fort St. James, the country was almost impassable, owing to the
" constant rains, but the soil is rich, and grass and weeds were very
" luxuriant. The country around Lakes Tsin-kut, Ta-chick and Nool-ki
" is very fertile, and from the occurrence of so much prairie, together
" with the similarity of the flora to that around Edmonton, I consider
" the climate of the two regions to be much alike. The former, though
" further north, is less elevated, and this, together with the well-known
" northern trend of the isothermal lines in N. W. America, more than
" compensate for the difference in latitude. The dry summer climate,
" which is indicated by the flora, proves the rainfall to be inconsiderable,

"and, therefore, the prospects are good for the successful cultivation of "grain."

Westward from that fertile piece of country, the northern line will open up 90 miles of the Wotsonqua Valley, which Mr. H. Cambie describes as such a fine pastoral region, and even fit for agriculture, but for the "*bug-bear*," the summer frosts, so that it can be claimed for the northern route, that it will open up as much fertile country on the Pacific slope, as its southern rival—which is not much to boast of.

It has been shown by the evidence of Mr. Fleming's engineers, that the Frazer and Thompson river valleys offer but a trifling quantity of agricultural land. The Reverend George Grant—Mr. Fleming's Secretary on his overland journey in 1872—tells us very plainly that British Columbia is *not* an agricultural country, that it is a *Sea of Mountains*, and ill-calculated to attract immigration, in fact, the exodus of whites during the last fifteen years, has greatly exceeded immigration during the same period. Let us now examine some further testimony bearing particularly upon the "rich lands," recently alleged by the *Mail* and *Globe* newspapers, to be available for settlement upon the Burrard Inlet route.

In the leading article of the *Mail* of 6th May last, an extract from which has already been given on the first page of this paper, we are told :—

"The main advantage in adopting the Burrard Inlet route, par-
"ticularly as opposed to the northern or Port Simpson route, has been,
"and will still further be found to be, in the expediency of the
"present route for colonization purposes, for opening up the best
"western lands, and for facilitating progress into the prairie country.
"The opening up of the Canóns of the Fraser to which Mr. Blake
"so strongly objects, will, it is alleged, on the fullest and best evidence,
"give up a large area of fine wheat lands to prospecting settlers.
* * * "There can be no doubt that the adoption of the Burrard
"Inlet route, for all the purposes of settlement, agriculture, trade, and
"the peaceful growth of a great region, has been wise."

The *Globe* ranted even more wildly during the last Session of Parliament, much to the delight and astonishment of the Frazer River partisans. Now, every sensible man felt at the time that those organs spoke rashly, and with but a glimmering of truth to bear out their assertions. Where are those fertile areas ? Whose evidence is the fullest and best? It must have been highly interesting to hear honorable members from the Mainland of the "Sea of Mountains," descant, no doubt most eloquently, upon the prospects of intending settlers. Why, upon the lower 85 miles of the sections now let to contract, *i.e.* between Yale and Spence's Bridge, there are not five hundred acres of arable, cultivable land upon which to put a plough or harrow. I shall here quote from Professor Macoun's Report of 1875-76, pages 114, 116, 120, 121, of the Geological Report of that date:—

" The valley of the lower Frazer, for agricultural purposes, may be
" said to end at Sumass; but there are numbers of small locations where
" farming could be done on a limited scale as far up as Fort Hope. Be-
" yond this point the valley becomes confined between the mountains,
" and these press so upon the river, that, before reaching Yale, the
" traveller realizes what a canôn is, and the mind is tortured with the
" thought of what might happen if anything went wrong with the boat
" or its machinery."

" The Lower Frazer valley has along its left or south bank a range
" of low rocky hills, extending from Langly to the mouth of the Sumass
" River; and to the southward of these, between them and the spur of
" the Cascades before mentioned, lies the Sumass prairie. Nearly in the
" middle of this prairie lies the lake of the same name, about ten miles
" long and four broad in its widest part. During the season of flood it
" extends from hill-foot to hill-foot, and even after the subsidence of the
" waters its mud banks or beaches reach certain points on both sides.
" The larger half of the prairie is at the south-west end of the lake, and
" is about four miles square."

" The prairie ground at the north-east end of the lake is bounded by
" a belt of trees, separating it from the clear or prairie ground on the
" banks of the Chilukweyuk River. The clear ground on both sides of
" this river has been apparently formed, partly by the repeated action of
" fires destroying the trees, which at one time grew on the higher banks,
" and partly by the action of the floods which annually submerge a large
" portion of it. These prairies have, during the season of flood, very
" much the appearance of immense lakes, being, with the exception of a
" higher ridge here and there, almost entirely covered by water. When
" the water subsides the growth on these low grounds and prairies is
" most astonishing, reminding one of the luxuriance of the tropics with-
" out its peculiar vegetation."

" On the afternoon of the 18th, I started on foot, expecting a con-
" veyance to overtake me and carry me to Boston Bar that evening. As
" I wended my way along the river, now examining a steep cliff or peer-
" ing down a chasm in search of cryptogams, the Indians would leave
" their fishing to look at me, but never addressing a word would gaze for
" a short time and disappear. On the dripping rocks along the road I
" obtained fine fruiting specimens of many mosses, prominent amongst

" which were *Bryum crudum* and *albicans*, and another unknown to me,
" *Polytrichum strictum* was in fine fruit, and various species of *Grimmia*.
" *Racomitrium, Mnium, Orthotrichum, Hypnum* and many others well re-
" paid me for my trouble. The *Alsia abietina* was very abundant at
" times, and the damp faces of many rocks were covered with beautiful
" Hepatiæ. The only flowering plants of any note were *Arnica cordifolia*
" and *Smilacina uniflora*, which were not uncommon. A few miles on
" the Yale side of Boston Bar we turned the point of the mountain, and
" almost immediately the plants showed a change in the quantity of
" moisture, and, on looking back, the eye at once detected the cause, in
" the mountains acting as a barrier to keep out the superabundant
" moisture of the Lower Frazer."

" Lytton is a poor, miserable place, only having three gardens in
" the whole village. By utilizing the small brook which comes from the
" mountains behind it, many fine vegetables could be raised, as the soil,
" where not too much encumbered with stones, is good. Between
" Jackass Mountain and Spence's Bridge *there is very little cultivable land*,
" and this requires to be irrigated before good crops can be raised."

Doctor Dawson, at page 246, Appendix S, (Mr. S. Fleming's Report of 1877) says of the extent of cultivable land in British Columbia :—

" It is very difficult, with the information now accessible, to form
" even an estimate of the quantity of arable land in the interior of British
" Columbia. I have only seen a few parts of the southern portion of the
" interior plateau, but judging from these, and facts obtained in other
" ways, I am inclined to believe that *the cultivable land east of the Frazer is*
" *probably in area less than 1,000 square miles.* It is to be remarked, how-
" ever, that this area does not at all adequately represent the capacity of
" the country to support a population, as a comparatively small patch of
" arable land serves the stock-farmer, whose cattle roam over the
" surrounding country. West of the Frazer, as far north as the Black-
" water, the cultivable areas are very small. The so-called *Chilicotin*
" *Plains* lie *too high for farming* and the available area in the valley of the
" Chilicotin was roughly estimated by me in my report for 1875, at *7,000*
" *acres only*. An area of 300 square miles might be perhaps taken as an
" estimate of the farming land of this region. North of the Blackwater
" is the *Lower Nechaco basin, already more than once referred to. The area*
" *of this is probably about 1,000 square miles.* Bordering on Francois
" Lake are considerable stretches of country not raised so much as 300
" feet above it, and, therefore, considerably below the 3,000 foot contour.
" The soil is very fertile, and the vegetation much resembles that of the
" *white silt* basin. Supposing this country to be suited to the growth of
" barley, oats and the hardier root crops, which appears highly probable,
" though no trials have of course been made, an area roughly computed
" at about 200 square miles will be added."

The " cultivable land *east* of the Frazer," referred to in the above quotation, as being, " in area less than 1,000 square miles," (about equal in extent to an average county in Ontario,) is the only *land* which will be opened for settlement, by the construction of the entire length of the British Columbian section, which is, by Mr. S. Fleming's estimate, to cost $32,000,000.

From the summit of Yellow Head Pass to the confluence of the Clearwater and North Thompson—180 miles—the railway line is within the Rocky Mountain ranges, and there is no land fit for cultivation. Thence down to the meeting of the two branches of the Thompson, the slopes of the hills are covered with "bunch" grass and groves of fir and aspen, but the valley in many places still partakes of the canon character, and, until within 15 miles of the junction at Kamloops, there is scarcely, if any, land fit for settlement, a total of 240 miles of perfectly barren country, devoid even of minerals, for gold prospectors have, from time to time, thoroughly examined it. The largest tract of arable land in the valley is contained within the angle formed by the two rivers, and is occupied by an Indian Mission.

At Kamloops the line may fairly be said to have reached the fertile zone lying between the Rocky and Cascade Mountains. This fertile zone is exceedingly limited as to extent, consisting principally of the interval land in the narrow valleys of the Thompson, Grand Prairie, Similkameen, Tulamene, Nicola, Buonaparte, Frazer, and lateral connecting valleys.

Nearly all the good lands are taken up by speculators, and but a small proportion is yet under cultivation. Nearly all those lands require irrigation, which, when obtainable, conduces to the production of abundant crops, as in the Utah Valley.

Although those arable lands vary in altitude from 1,000 feet to 2,000 feet above the sea, they do not suffer materially from summer frosts.

Such, in brief are the " fine wheat lands " which the construction of the most formidable 125 miles of railway in the world is to " open out for prospecting settlers." And here, let it be understood, that the localities just named by no means embrace the area of 1,000 square miles alluded to by Doctor Dawson ; they only form a portion of that area, the balance being in the vicinity of Lac La Hache, Quesnelle, and along the upper portion of the waggon-road, and along the Frazer River, from Soda Creek upwards.

The grazing lands in the Kamloops section are, however, excellent, but not inexhaustible, for bunch grass when closely cropped gives place to sage and wormwood.

Mr. Sproat, a provincial authority, estimated the live stock in the Province in 1875 as follows :—35,000 horned cattle ; 6,000 to 7,000 horses ; 12,000 to 15,000 sheep ; 10,000 hogs.

The total white population within this district, *i.e.*, from Yale upwards and eastward, that is to say, as far as Cariboo to the north, and Kootenay to the east, according to the Directory of 1874, was about 1,400, distributed as follows :—

Yale	60
On waggon-road between Yale and Lytton	25
Lytton	42
On waggon-road, Lytton to Ashcroft	20
Ashcroft	6
Cache Creek and vicinity	40
Clinton, see Lilloet Clinton District	00
On road, Clinton to Lac la Hache	10
Lac la Hache	20
Williams Lake and St. Joseph Mission	11
Deep Creek, Soda Creek and Alexandria	25
Quesnelle	60
River Trail, Williams Lake to Lilloet	22
Cariboo	524
Lilloet, Clinton District	250
Thompson River, Nicola Valley, Kamloops	170
Kootenay	108
Similkameen and Okanagan	30
Total	1,423

On the line of railway under contract, Yale to Kamloops (125 miles) (Directory, 1874), the white population is as under :—

Yale	60	
On waggon-road, Yale to Lytton, 56 miles	25	*Overestimated.*
Lytton	42	
On waggon-road, Lytton to Ashcroft	20	"
Ashcroft	6	
Cache Creek and vicinity	40	"
Thompson River and tributaries, including Kamloops	150	
Total	343	

It will be said that since 1874 the population has increased. It is true that since that period a few individuals have found their way into the interior of the Province, but the number has been so limited as to scarcely merit consideration, and, as the population given in the foregoing lists actually exceeds by several dozens the figures in the Directory, we may assume, with tolerable certainty, the actual present population to be but slightly in excess of the number given. There has been a slight increase of population in the New Westminster District during late years. The population of the Frazer Valley *below* Yale will be considered further on.

A railroad built upon the verge of a precipice for a distance of 75 miles, through mountain gorges in which no settlement can ever take place, is not what is required to open communication between the excessively sparse population of the interior of British Columbia and the seaboard. Such a road, once built, would do no business. One passenger train and half a dozen freight trains would transport the entire population, bag and baggage, farm produce and all, and then, what would remain ? Two streaks of rust and the right of way—a monument to Canadian folly.

Already there is in those canôns as good a waggon-road as need be. A foot passenger upon that road may walk mile after mile without meeting a team. In fact, the present traffic could be increased fifty - fold without sensible inconvenience, therefore, the construction of a railway between Yale and Kamloops is such an absurdity, that one may well pause to wonder at such a proposed waste of money. In justice, however, let it be said that this matter has never yet been brought home to the full understanding of the masses directly interested. The people of Canada have been obliged to glean the little they know of this subject from unreliable newspaper articles, and from the official reports of the Railway Department, neither unimpeachable sources ; and there are so many private interests involved, so many speculations depending for success upon the inauguration (though not necessarily upon the consummation) of this project, that it has well nigh become impossible to arrive at the truth ; in fine, no subject has been more discussed, less understood, and more impudently misstated than this. But the warning may be too late, and Canadians, as each year of woeful waste rolls on, with ever increasing taxes, and an unbearable public debt, must resign themselves to the inevitable and calmly await the financial ruin and political disintegration, which must result from such a railway policy.

Below Yale, the head of navigation on the Frazer River, and west of the Cascade Mountains, the total quantity of land fit for settlement eventually, has been estimated at 520,000 acres. How much there is of good land worth cultivating has not been yet ascertained with certainty, but it is estimated that about 10,000 acres are under-cultivation, or more strictly speaking, under occupation, at the present time. Those lands are favourably situated, being pierced and partially surrounded by navigable waters. They are accessible all the year round from Victoria, or the coal mines of the Nanaimo District, where farm produce is in constant demand. Notwithstanding all those natural advantages, it is a strange fact that scarcely a bag of flour has ever been exported from the Frazer Valley. On the contrary, flour from San Francisco or Portland in Oregon, and bacon from Chicago is, or has until very recently, been imported for use in the interior. The cause of this is, doubtless, to be ascribed to the almost periodical inundation of this fine land by the high floods of the Frazer in Spring.

Mr. Marcus Smith in speaking of this District, at page 45, Report of 1878, says :—

" Below Hope the valley begins to open up, and it becomes several
" miles wide, in places, before New Westminster is reached. The bottom
" flats are generally low and partly prairie land ; the river meandering
" through them is occasionally divided into channels or sloughs, forming
" numerous islands; these are thickly clothed with cotton-wood, vine,
" maple, willow and other woods. There is good land on the higher
" benches, though but little wheat is grown in the district. The reasons
" for this, as given by the farmers, are :—The uncertainty of the weather
" during the harvest season, the alternate rains and hot sunshine causing
" the grain to grow in the ear before it can be housed ; and, further,
" that they find it more profitable to raise stock, coarse grains, hay, and
" fruit, and import their flour than to spend money in producing wheat,
" which, at best, would prove to be but an inferior article. The cattle
" are reared for the markets of New Westminster and Victoria ; the hay
" and oats are sent to the logging camps, and the fruit to the upper
" country.

" The total area of land in the valley is estimated at a little over
" 500,000 acres : of this but a very small part is under cultivation, and
" it will require much labour and expense before any extensive increase
" can be obtained. The great bulk of the land that could be most
" easily brought under cultivation, lies on the estuary of the river below

"the point where the line leaves the valley for Burrard Inlet; and most
" of the balance is on the opposite side of the river to that on which the
" line is located. Much of this land is subject to overflow from the
" floods of the river and from high tides in the Strait.

"Taken altogether, this is a very fine district, and in course of time
" will have a considerable population; but it is obvious that the recla-
" mation of the low lying lands is not to be brought about by a railway,
" but by means of dykes, embankments, pumping machinery and such
" other works and appliances as have been successfully used on lands in
" a similar condition.

" Steamboats already ply between New Westminster and Yale (90
" miles) twice a week each way, and would do so daily if there were
" sufficient traffic. These steamers stop at any point on the river where
" desired for the collection of passengers or freight, however limited in
" number or quantity; a degree of accommodation greater than could be
" afforded by any railway. The amount of traffic which the valley would
" supply to a railway would be but limited, as its main products go sea-
" wards, and four-fifths of the traffic, both of passengers and freight,
" which passes up into the interior is in connection with the Cariboo
" Gold Mines, for the necessities of whose development there must, and
" will ultimately, be found a shorter and better route from some point
" on the coast further north. On the whole, it does not appear that the
" prospects of a railway on this route are encouraging."

According to the certified list of 1876, *the number of voters* in this district was *581.*

From the Victoria Directory of 1874 we gather that there were at—

Burrard Inlet	156	Whites.
New Westminster	164	"
North Arm Frazer River	21	"
South Arm " "	80	"
Matsqui	22	"
Sumass	89	"
Chilliwack	64	"
On Frazer above New Westminster	29	"
Langley	46	"
Boundary Bay and Semiahoo	21	"
Hope	29	"
Total	621	

A Memorandum taken from the Census of 1871 states:—

" The District of New Westminster (see also Burrard Inlet and
" Frazer Mouth lists,) returns a population of 1,292 whites, 27 Chinese,
" 37 coloured; natives, no returns, say 300. Total, with natives,
" 1,650."

It thus appears that the population of 1871 exceeded that of 1874. The lists may be erroneous, in any case, the population is exceedingly scanty.

As Mr. Smith truly says, it is not a railway that is required to bring about prosperity to this district, but dykes, pumping machinery, etc., and, in the central plateau, branch roads, the improvement of the waggon road, and such public works as are, in actual justice, required for such a sparse population and limited area of agricultural lands. To this end it is not necessary to build a railway 125 miles in length, costing $12,000,000.

As has been remarked, on a preceding page, half the mileage between Yale and Port Moody is classified by the Engineers as heavy. The distance is 90 miles, alongside the navigable waters of the Frazer; of course, the intention is ultimately to carry the railway to Burrard Inlet, so to complete the grand trans-continental route, and build up a great city at Burrard Inlet, if possible.

From a point on the Frazer River in the vicinity of Sumass, the distance to Coal Harbour, Burrard Inlet, is about 40 miles. The works will be very heavy along some portions of this piece of road. From Coal Harbour to Cape Flattery the distance is 150 miles, including some very intricate and dangerous navigation according to the authority of Commander Pender, who, at page 300 of S. Fleming's Report for 1877, says:— " For reasons given in No. 27, Burrard Inlet is, in my opinion, preferable to either of the other places named; but even here the risks attending the navigation of large steamships, against time, amongst the islands lying between Fuca Straits and the Strait of Georgia, are, to me, very great."

Other naval authorities admit that the approaches to Burrard Inlet from the Straits of Fuca involve more or less intricate navigation, and that the San Juan group of islands, commands those approaches.

Admiral DeHorsey says :—" The tortuous channel from Burrard
" Inlet to sea, through Haro Strait, will frequently be unsafe on account
" of the strength of the tide, great prevalence of fog and absence of
" anchoring depth. Burrard Inlet itself also, although possessing a safe
" port in Coal Harbour, and a good anchorage in English Bay has these
" objections, viz. :—that the narrow entrance to Coal Harbour through
" the First Narrows is hardly safe for large steamers, in consequence of
" the rapidity of the tide, and that English Bay, although affording
" good anchorage, would not, in my opinion, be smooth enough during

"north-westerly gales for ships to lie at wharves, there being a drift of forty miles from the North-West."

And with regard to fogs, many navigators have stated their belief, that there is a greater frequency of dense fogs in the Georgian Gulf than on the northern coast; in this connection, I may not inaptly quote from page 108 of S. Fleming's latest report, in which Doctor Dawson writes:—

"I have elsewhere stated that fogs do not seem to occur with such frequency in the vicinity of the Queen Charlotte Islands as in the southern part of the Strait of Georgia. La Perouse, the great, but unfortunate navigator, wrote: 'I first thought these seas more foggy than those which separate Europe and America, but I should have been greatly mistaken to have irrevocably embraced this opinion. The fogs of Nova Scotia, Newfoundland and Hudson's Bay have an incontestable claim to pre-eminence from their constant density.'"

Captain John Devereux says, at page 308, Railway Report of 1877:—

"Burrard Inlet has a safe and commodious anchorage, two (2) miles inside the first narrows at Coal Harbour, also another seven (7) miles inside the second narrows at Port Moody; but, there is one great objection to either of these places, viz.: both the first and the second narrows, respectively, are, but about a cable and a-half wide, through which the tide runs about nine knots an hour, creating whirls and eddies, rendering it unsafe for large steamers to enter or leave port at night, or at certain stages of the tide, leaving out all interruption by fogs and thick weather, which occur more frequently inside than out."

"At English Bay, at a place marked on the chart as Government reserve, is a good anchorage, with every facility to construct a breakwater and wharves, and by erecting a light-house on Passage Island, one on East Point, one on Twin Point and another on Discovery Island, the largest ships might be conducted thither in safety; but there are three months in the year, viz.: from part of August to part of November, when this coast is subject to dense fogs, rendering it unsafe, if not utterly impossible, to navigate Haro Strait and the Gulf of Georgia with large steamers, such as the Royal Mail, Cunard and Pacific Mail Co.'s ships."

"This will, I think, be conceded by all who know anything about such ships and the straits in question, where the tide runs from four to six knots per hour, with boiling rips and overfalls, narrow channels and outlying reefs. * * * The fogs are so dense here that land cannot be seen one hundred yards off."

From the same point on the Frazer, near Sumass, a railway can be built without much trouble and at little cost, through a flat region, to a point on Puget Sound known as "Holme's Harbour." The length of this line would be about 66 miles. It would cost very much less than that portion of the Canadian road terminating at Coal Harbour, and the terminus would be at a magnificent harbour within easy distance of Cape Flattery, say 85 miles. The navigation is unparalleled, being perfectly free from danger, and ships can reach this point without towage. This harbour is situated under the lee of Whidby Island, U. S. Territory, and the intention is to cut a canal two miles in length across the neck of land separating this port from the waters of Admiralty Inlet. The citizens of the United States are quite alive to the importance of the matter, and regard this place as the natural outlet of Canadian Pacific traffic, *via* the Frazer River. And they are perfectly right. Freight will follow the most economical route. From Sumass to Cape Flattery, *via* Holme's Harbour, the distance will be 151 miles. Between the same points, *via* the Burrard route, the distance is 190 miles. In the former case, rai freights will be lower than on the expensive road to Coal Harbour, while rates of insurance, towage and pilotage will be very much less than on Canadian waters. The fact is undeniable, dispute it who may. Mr. Marcus Smith has already pointed it out in his Report of 1878.

But if further testimony be desirable, we have but to glance over Mr. S. Fleming's Report of April, 1880, wherein, at page 146, Major-General Moody, formerly Commander of the Royal Engineers in British Columbia, gives his. He is a strong partizan of the Frazer River line, and in a lengthy paper upon the railway question shows his clear perception of the inevitable tendency British and American commercial relations will have to co-mingle, and trade to gravitate towards the most favourable outlet.

" One must keep in mind that if Route III did not exist, the
" material interests, present and future, of this valuable south portion of
" British Columbia, from the seaboard to the Rocky Mountain range,
" would gravitate inevitably to the foreign branch lines of the United
" States' North Pacific Railway; such branches coming up from south
" to different points along the frontier, east and west of Cascade Range."

" The coast branch up from the future *great and important port of*
" '*Holme's Harbour*,' (U.S.,) in the Straits of Georgia, to Semiahmoo
" Port (U.S.,) 45 to 50 miles, will reach to about 15 miles from New
" Westminster, and, as a matter of course, in the progressive inter-
" change of trade and communications between the two nations, will
" extend to New Westminster.

3

" Another branch will also probably reach *a point higher up the Frazer*, nearer Hope."

His argument, it must be borne in mind, is to show the advantage of the Frazer River line. Were no such road ever to come into operation, British American commercial relations would be more tightly banded together, and despite the " great strain on the sense of duty," the mixed population (however loyal) would go over to the enemy. He adds :—

" Any results as above would not only be effectually counteracted by line Route III., but, as before stated, *additional gain may be looked for from over the border.*"

In any case, line or no line, the Americans will run a branch of the " North Pacific " up to the frontier. Such is his deduction.

The probability is that the " Canadian Pacific," if ever built on the Frazer, will never go beyond Sumass, for the reasons just given. The "Grand Trunk" is similarly situated; at both ends long lengths of the line run upon United States soil, and for this deplorable condition of affairs we must blame the idiots who were intrusted with a duty utterly beyond their limited powers. As a matter of fact, every United States citizen who visits British Columbia, sees the country and investigates the railway question, goes home with the idea firmly impressed upon his mind, either that Canadians have too much money, or that their rulers are, to put it mildly, greatly in advance of the age. They laugh to scorn the idea of a " canôn railway."

At present the construction of the Pacific Railway Western Section is altogether premature. It should, and probably will, when Ontario and the Eastern Provinces wake up to a true conception of the gigantic fraud being perpetrated upon them, be deferred indefinitely. In any case, not a sod should be turned upon the western slope of the Rocky Mountains until the prairie section shall have reached the eastern confines of the Pacific Province. Until then, other works of more vital importance to the Province than the gigantic and expensive toy the British Columbians have been taught to hanker after, should be entered upon with as much liberality as may be commensurate with the length of Canada's purse strings. Every nerve should be strained to preserve amicable relations with the Pacific Province, but, to this end : not a cent, beyond what strict justice to the older Provinces dictates, should be abstracted from the pockets of the people.

A careful perusal of the foregoing pages must convince the most sceptical reader that the truth, and nothing but the truth, has been the writer's aim throughout. The battle of the routes has been waged at such odds, so many tongues have been tied, so many valuable reports suppppressed, that the public has not hitherto been in a position to form any opinion as to their respective merits. No sane reader can entertain further doubt with regard to the preponderating merits (climatic and agricultural) of the country between Lac La Biche and the Pine River Pass, over that of the region west of Edmonton, upon the southern line. The best authorities have been appealed to, and they have spoken. With regard to that part of the southern line between Livingston and Edmonton, the following extracts from Surveyors' Reports (already given by a previous writer upon the same subject) will serve to convey the truth:—Those extracts are taken from the reports of the Department of the Interior.

" To Fishing Lake (long. 103¼) a distance of 19 miles. The soil " throughout is good sandy loam, and most of the timber of useful " dimensions."

" To Big Quill Lake (long. 104¼) a distance of 32 miles. Well " supplied with wood and water, having a soil sandy loam of fair quality, " lying between Quill Lake and the Touchwood Hills. The streams " running into Quill Lake are fresh, whereas the lakes are strongly " alkaline."

Turning northward for 20 miles, to a point beyond the railway, the surveyor's line is reported thus :—

" The first six miles are on the sandy alkaline strip between Big " and Little Quill Lake. Some fair sized timber is found here, but the " soil is poor ; and continues so through a more open country, until " within 3 miles of the C. P. R. line, when we encounter rising ground, " densely-wooded, with large poplar and numerous ponds."

Turning westward on the 10th base, the survey proceeded at an average distance of about ten miles from the railway for a stretch of 180 miles. The Report of 1877, says of that line :—

" The wooded and pond-country continues for about 27 miles, when " the country becomes more open and inviting; and continues so to the " 40th mile, when we gradually descend into an almost barren, rolling, " alkaline, sandy plain.   *   *   *   For about 24 miles the " line runs through the same sandy, rolling plain. On the 13th mile " we crossed the Canadian Pacific Railway line, where it deflects to the " north, 2 miles south of an alkaline lake."

The Report of 1878 continues to review the same survey line from the close of the above. It says:—

"I experienced great difficulty in making progress (for 108 miles) on the 10th base, owing to the want of wood and water, the country along that line being almost destitute of both. On one section of it water had to be carried for the party, and wood for posts and fuel, in our carts, for a distance of 32 miles. The soil on the part surveyed of this line (108 miles), with the exception of some few miles in the Eagle Hills, is of a poor nature, being light and sandy, and in most cases alkaline. In fact, none of the country between the 106th meridian and the point at which I turned northward (an interval of over one hundred miles), is of any use for agricultural purposes."

Turning northward at the end of the line just reviewed, the surveyor describes the country traversed (for 36 miles) thus:—

"Of a better nature than on the 10th base; for though the soil is light, it is well watered, and the pasturage is excellent. It is, however, destitute of wood."

From Battleford to the 110th meridian, the line (75 miles in length) is reported thus:—

"The soil, generally, is exceedingly poor; and although improving a little in the immediate vicinity of Battleford, is even there very light and sandy. * * * From the Meridian Ranges 18 and 19 to the 110th Meridian, the country is decidedly more attractive. For the first 30 miles there is a scarcity of wood, but water abounds. Indeed as a rule this was the only country (in a course of over 300 miles) passed over, in which the water met with was not more or less alkaline. * * * From the exceeding richness of its grasses, and the special fitness of the kinds produced, I am led to believe that it (a tract of 30 miles wide near the 110th meridian) excels as a grazing country, anything I have seen in Manitoba or the North-West Territories."

"In summary of the foregoing and of other evidence on the subject, it may be concluded that for 300 miles across the plains the adopted route, while presenting exceptions here and there, traverses a region whose soils and other circumstances may be said in general to be unsuited for agricultural settlement."

Mr. M. Aldous, of the western special survey, says, of that portion of the country between Fort Pitt and Edmonton, at page 41, part II., (1879), Report of the Department of the Interior:—

"In the whole distance surveyed between the 110th and 114th meridians, we have not passed over a single mile of what I deem

" worthless land. * * * The streams contain good clear
" water, and but few of the lakes or ponds are alkaline."

At page 45 of the same Report, Mr. A.P. Patrick, D. L. S., thus speaks of the country south from Battleford':—

"I left Battleford for the Forks of the Red Deer and South Sas-
" katchewan Rivers on the 6th August, 1878. The country passed over
" for the first thirty miles may be said to be fit for settlement, though
" the soil is light and wood scarce. From this point to the Forks (138
" miles) the soil is fair but dry, and in my opinion unfit for farming, no
" wood, and water only to be found at great distances."

Mr. H. MacLeod, the Engineer who has been in charge of the Railway Surveys upon the prairie section, estimates the proportion of poor soil between Winnipeg and Lake St. Ann, to be nearly one-half. *Vide* Report of 1877.

With the exception then of the country between the Meridians of Fort Pitt and Lake St. Ann (225 miles) the region traversed by the now located Canadian Pacific Railway west from Livingstone, is of but a medium character, and very much of it quite open or treeless.

On the other hand, by the Northern or Pine Pass line. branching northwestward from Livingston *via* Fort à la Corne towards the Beaver River country and Lac la Biche, (some 470 miles) the country is nearly all fit for settlement. (*Vide* Surveyor-General's, Mr. Macoun's, Mr. Smith's, and W. F. King's reports.)

It may be as well to remind the reader that the above opinions were expressed by the writer several years ago. (*Vide* " Canada on the Pacific.)

It must strike every intelligent observer that, without actual surveys beyond the explorations in a general way which have been made of late years throughout the North-West, no very reliable estimates of the quantity of arable land available for settlement can be expected.

In 1872, Mr. Macoun accompanied the writer through a portion of the Peace River country. In 1875, Messrs. Selwyn and Macoun again saw a portion of the same region; and in 1879, Doctor Dawson and his assistant had opportunities of examining very much more of its southern portion than any previous explorer. The last named gentleman is, therefore for that reason alone, if for no other, in a better position than any one else to make statements as to the soil, climate, and available areas of that beautiful country.

But a hasty journey through an Indian country, under the unavoidable difficulties (trifling, no doubt, but harassing withal) which constantly beset the traveller, cannot be regarded as the basis of dogmatical assertions as to the areas, arable or non-arable, available for settlement. It may be permitted to a traveller, under such circumstances, to hazard an opinion, a guess, but such statements should be accepted by the public, *cum grano salis.*

Mr. Dawson is, however, very guarded in his statements, takes care to impress upon his readers the generality of his views, and abstains from magisterial assertion.

Mr. Macoun, on the other hand, an ardent admirer of nature and a zealous botanist, has allowed himself a latitude of expression with regard to *areas* suitable for agriculture in the North-West, which, to say the least, can only be regarded as ideal.

In Mr. S. Fleming's Report for 1877, at page 336, Mr. Macoun has made a classification of the lands in the Canadian North-West. Taking for example, of the five acres described, that of the Peace River country, it will be seen that he estimates the available quantity of arable land to be sixteen millions acres.

It may be interesting to know by what process he arrived at the above result. Referring to the Geological Report of 1875-76, it appears he descended the Peace River by canoe to Fort St. John, thence ascended the northern slopes of the river valley, walked northward a distance of nine miles, and returned over the same trail to his starting point.

From Fort St. John he descended by raft or canoe to Dunvegan, thence by canoe to Fort Chippewyan and Athabasca. He landed only to samp, and also made botanical examinations at Battle River, Vermillion and Little Red River. Upon two or three occasions he penetrated the country to a distance of half a mile from the Peace River. From Athabasca the Hudson Bay Company forwarded him *via* Portage La Loche (Methy Portage), Isle à la Crosse and Green Lake, to Carlton. The latter portion of his journey was also by water, with the exception of some 140 miles between Green Lake and Carlton.

His journey from Dunvegan to Carlton (1,100 miles) occupied two months. All his movements were hurried, owing to lack of provisions while descending the placid Peace River, and on the Isle à la Crosse route, from the fact that he was a passenger in the Hudson Bay boats. (*Vide* pages 156 to 165 of his report.)

Between Dunvegan and Fort Chippewyan the Peace River flows at the bottom of a valley which decreases in depth, from 700 feet at the former, to 50 feet and less at the latter place. It was consequently impossible to see anything of the surrounding country from a canoe.

How then, in the name of common sense, can he justify his sweeping assertion that there are sixteen millions acres of arable land within the section of country drained by the Peace River, east of the Rocky Mountains?

Does he think that British capitalists will swallow such an unfounded statement?

To take a map and measure off certain unknown and unseen areas finished this remarkably easy method of "doing" the country.

That there are vast areas suitable for settlement there is every reason to believe, but there is no justification for deliberately misleading the public with an array of imaginary figures. Facts, not fancies, are wanted.

Similar wild estimates have, perhaps, been made in other parts of the North-West, and the writer is not alone in decrying such a wholesale method of survey, for, in the early part of 1879, when he brought the impropriety and absurdity of jumping areas in this manner, under the notice of the Minister of Railways, and of the Chief Engineer, Mr. Fleming, the latter quite concurred.

Doctor Dawson's examination of the southern portion of the Peace River country during the season of 1879 has however removed in great measure, any doubts as to its value and extent.

In 1872, the writer hazarded the opinion (see "Canada on the Pacific") that, in the southern Peace River country there would *probably* be found available for settlement, agricultural land equal in extent to the original Province of Manitoba. This view has been more than justified by Doctor Dawson.

In the last report issued, Mr. Macoun exhibits a map "indicating the limits within which good lands are known to exist, west of the 101st meridian."

On the western portion of that map there is a tract or triangular area of about 10,000 square miles, shown as prairie and good land, which is well known to be forest and worthless for agriculture, i.e., that piece of country extending from Rocky Mountain House northward to the River Pembina, and thence west and north to the Rocky Mountains. The Southern (Burrard Inlet) line is projected through this worthless country, which Mr. Macoun speaks of in his Report of 1877, page 828, as "*seemingly the worst part of the swampy region* near the Rocky Mountains."

Mr. Macoun's instructions were: To explore 60,000 square miles of the country west of Livingston and north of the 51st parallel of latitude.

He was probably five months in the field. Admitting that he travelled continuously during the whole of that period, at the average rate of 20 miles per day and that he was able to determine the quality of the soil for two miles on each side of his track as he went along, he would be in a position to report upon 12,000 square miles = 7,680,000 acres. He, however, affirms that there are 134,000,000 acres of good land between Manitoba and the Rocky Mountains, *exclusive* of the Peace River country.

How does he know this? He did not see one-twentieth part of this enormous area. His map is divided into sections represented by the parallelograms contained between adjacent parallels and meridians. He rode across (directly or diagonally) some 23 of those sections. Each of these sections represents an area equal to two counties of the Province of Ontario. Is it reasonable to believe that a hasty ride across an Ontario county would enable anyone to state its agricultural value?

Mr. Macoun states that much of the prairie country south of the 52nd parallel is better than has been reported by Palliser's expedition. It may be interesting to quote from M. Bourgeau's botanical report to Sir William Hooker in 1858. This was a French gentleman of high scientific attainments, who passed two or three consecutive years in the Saskatchewan and prairie country of the Canadian North-West.

Page 246, Capt. Palliser's Expedition: " On the prairies. . . .
" As the country towards the south merges more into open prairie, the
" clumps of young poplars are found only nestling on northern exposures."
" The last outliers of the woods to the south generally consist of 'islands' as

"they are called, which make a show from a distance, but when approached are found to consist of a small species of willow."

"The true arid district, which occupies much of the country along the South Saskatchewan, and reaches as far north as latitude 52°, has, even early in the season, a dry, parched look. In the northern district the accumulation of *humus* and the distribution of the pleistocene deposits has given rise to a variety in the nature of the soil; but to the south the cretaceous and tertiary strata almost everywhere come to the surface, so that the stiff clay, highly impregnated with sulphates, bakes under the influence of the clear sun of early spring, into a hard and cracked surface, that resists the germination of seeds. This must be the principal reason for the arid plains ranging to such a high latitude, as there is quite a sufficient quantity of moisture in the atmosphere during the summer months, to support a more vigorous vegetation, as is shown as far south as latitude 49° 30' N. when at the Cypress Hills, south sides of deep river valleys, and other expanses sheltered from the sun's rays until he acquires a considerable altitude, are found to be covered with pines, spruce firs, poplars, and abundant varieties of the vegetation found further to the north."

"In the arid plains, the plants, most evidently different from those regions to the north, are small *opuntias*, also the *sage* of the Americans."

"Much of the arid country is occupied by tracts of loose sand, which is constantly on the move before the prevailing winds."

"This district, although there are fertile spots throughout its extent can never be of much advantage to us as a possession."

"Along the base of the Rocky Mountains there is much fine land, with very rich pasturage."

Mr. Bourgeau, a most able botanist, passed a very much longer period in the North-West than Mr. Macoun, but a glance at his report will show that he never went so far as to classify areas. He quite admits the existence of vast tracts of excellent land south of the North Saskatchewan. He speaks highly of the agricultural capabilities of the country between Carlton and Edmonton, mentions the navigability of the two Saskatchewans and their largest tributaries, and in fine, gives a report which, from an economic and scientific point of view, would be difficult to subvert.

It is, however, quite apparent that the botanical testimony clashes in some important particulars. The public may judge from the statements made, which of the two is better entitled to credence.

The brown line drawn from Cumberland House to lesser Slave Lake, upon Mr. Macoun's map, page 245 of report, shows the northern limit of the area within which good land abounds.

Does he retract his statements of 1875, regarding the lower Athabasca? *Vide* Geological Report 1875-76, page 170.

In Appendix 14, Mr. Macoun, speaking ot the aridity characteristic of large tracts of country north of the 49th parallel, and south of the north Saskatchewan, refers to the cretaceous clays as the cause. He also mentions that the breaking up of the soil assists growth, besides some other well-known facts.

This is quite true, but he should have mentioned that those facts and theories have been long well-known. He has culled his information from Doctor Dawson's report in connection with the Boundary Commission, and from other well-known works published some years ago.

This cretaceous formation, which Mr. Macoun admits to be the cause of sterility, is fully discussed in Doctor Dawson's Report of 1875.

"347. *Fort Pierre Groupe.* This group appears to occupy a very
" great extent of country in the region north of the 49th parallel."
* * *

" 352. Dr. Hayden writes : ' This formation is the most important
" one in the cretaçeous system of the North-West.' " * * *

" Wherever this deposit prevails, it renders the country more com-
" pletely sterile than any other geological formation I have seen in the
" North-West."

" The contrast between the country resting on this formation and
" that based on the Lignite Tertiary, is very striking, and where the
" dry uplands of the Tertiary would seem, at first, less favourable than
" the low-lying plains of No. 4 ; the former can support a short thick
" growth of nutritious grasses, where the latter has the character above
" described."

Doctor Dawson defines the boundaries of the cretaceous subdivision No. 4, within British territory, at page 149 of his Report in connection with the Boundary Commission, and further on supplies abundance of information in regard to water supply, the climate, tree-growth and areas fit for settlement in the North-West. Mr. Macoun has dipped into all this, hides the fact, and presents his borrowed knowledge to the public as his own. This is not fair, either to the authors quoted or to the public, and will but serve to throw discredit upon himself.

Since writing the above, Mr. Macoun has gone forth in search of more acres. Upon this occasion his mission is to the south-west. It will be interesting to hear the result of this year's expedition. It cannot, however, be doubted that the Dominion will be further enriched by many more millions of acres. It may be taken for granted that another scientific adjustment of the map will be in order, and that much of the arid, cactus region north of the boundary line will be forever obliterated to make room for countless prospective homesteads. Plethoric capitalists will look forward with anxiety for the next Report.

In the summer of 1871 the first engineering parties from Ottawa were sent out, east and west, north and south. The writer accompanied the first prairie expedition under Mr. F. Moberly, and travelled from Fort Garry to Edmonton, Rocky Mountain House, the Kootenay plains, near Howe's Pass, back to Edmonton, thence to Jasper House, and back to Ottawa, during the period between August, 1871, and March, 1872.

In August, 1872, Mr. S. Fleming started from Fort Garry, Red River, with the avowed purpose of going over the line of route examined in the preceding year, to Victoria, B. C. The writer's services, as one of the members of the expedition of 1871, were called for to guide the Chief Engineer across the prairie section. The proceedings of the Chief Engineer's party upon that occasion have been duly chronicled in "Ocean to Ocean," a publication which, as its reverend writer remarks incidentally, purports to be "a truthful narrative."

At page 3 of the Report of the Engineer-in-Chief, dated 8th April, 1880, Mr. Fleming says:—" The first examination under my direction "was made in 1872, when I passed over the line from Lake Superior to "the Pacific." This sentence is scarcely correct, the last postulate being positively misleading. Mr. Fleming's expedition, consisting of himself, a clergyman, a doctor, Mr. Fleming's son, Mr. Macoun and the writer, travelled at the rate of 40 miles per day, between Fort Garry and Edmonton, over one of the many cart trails which intersect the country, but, far from following the then proposed, and now located, railway line, saw actually nothing of it, being at times from 70 to 100 miles to the north or south, according to the sinuosities of the trail. The expedition was, in fact, to all intents and purposes, under the control of the Reverend

Geo. Grant, who, from the very beginning, made strenuous efforts to
" run " the whole affair, as fast as possible, being, as he said himself,
excessively anxious to rejoin his parishioners at Halifax, by the 15th
November following. Accordingly, the examination of the prairie section
was to this end sacrificed.

At Edmonton, the party was broken up, the botanist and writer
going towards the Peace River, the others continuing on *via* the Hudson
Bay trail to Jasper House, and ultimately to Victoria, which they reached
early in the autumn. For particulars of what they did and saw, during
that memorable journey, the reader must refer to the volume alluded to,
" Ocean to Ocean." The writer of this paper, after passing the Rocky
Mountains, by the Peace River Pass, and sending the botanist home *via*
the Fraser River, finally crossed British Columbia, on snowshoes from
Fort McLeod to the Skeena and Naas Rivers, and reached the coast at
Fort Simpson in January, '73.

During the journey from Edmonton to McLeod Lake, *via* the Peace
River country, the writer being, from his experience of the country
between Edmonton and Jasper House in the previous year, well qualified
to institute comparisons, saw the probable advantages of the Peace
River route, or, more correctly speaking, of the *Pine River* route, over
the southern line, as means of access to Bute Inlet, that place being
then one of the termini most highly thought of. He accordingly reported
in favour of the Pine River route, in preference to that of the Peace
River, a proposition which created some disgust, and caused much
obloquy to be cast upon his judgment.

It has since been admitted by some of Mr. Fleming's engineers, who
are still staunch adherents to the "Yellow Head." route, that, had Bute
Inlet been finally adopted as the western terminus, the Pine Pass would
have offered the best route to it.

It has however, required many years to fully realize this, but the
final rejection of Bute Inlet a couple of years ago, paved the way for the
admission.

Until 1875, the writer favoured Bute Inlet as a terminus, but,
having since seen it, and the line leading to it for at least 150 miles, he
has been gradually convinced of its unsuitableness, especially within the
last year. In 1874, the writer was commissioned by the Government to
examine the Cascade Range from sea level to summit, between the

parallels of 52° and 54° north latitude. The sloop "Triumph" of the Geological Survey was, for this purpose, placed at his disposal, an examination of the various inlets made, and the result duly reported to the Chief Engineer. (*Vide* Report of 1877, page 137.)

This report, before its incorporation with Mr. Fleming's general report was, however strangely mutilated, and the portion treating of the coast from Douglas Channel southward to Queen Charlotte Sound, entirely suppressed.

This has lately proved to be a very unfortunate circumstance, as the matter of the suppressed portion entirely escaped the writer's memory until last winter, when an examination of Mr. Keefer's work on the Skeena between Fort Simpson and Kitsumkallum River led to a retrospect of the work of 1874.

Upon referring to the partially suppressed report of 1874, the writer found, at page 91 of his original MSS., the following passage :—

" It is needless to lengthen this report by more than a passing " allusion to the Kitimat Inlet, a huge water-filled indentation like the " others of the coast ; and, as there appears to be no passage from it to " the *interior plateau*, further reference to it here would be superfluous."*

But appended to this report, and marked for interpolation after the last passage, appeared the following remark :—

" At the north-east corner of this arm of the sea, there is a long " and narrow bay, which, were it dredged, would form an excellent " harbour. There is ample room for wharfage, but to deepen this bay, " the Kitimat, or at least one of its outlets, would require to be diverted " to the west side of the Inlet. A micrometrical survey has been made " by Mr. Richardson, during my absence in the interior while searching " for passes. Had I been successful in this respect, soundings of the " upper end of the Inlet would have been taken, and, in fact, a hydro- " graphical examination would have been made. As an outlet from the " upper Skeena, through the Cascades, the Kitimat Valley, apparently, " offers facilities unparalleled elsewhere on the coast."

The report was mutilated in four other places besides. The writer objected, but was told that the document was already too lengthy.

In the year 1877, Mr. H. J. Cambie was sent by Mr. Fleming to examine the Skeena and Wotsonqua Valleys, in connection with a line from Port Simpson to the interior. When at Kitsumkallum river, he

---

NOTE.—By "*interior plateau*" was meant the lake region immediately *east* from the Valley of the Kitimat and behind the "Cascades." The writer's instructions were to search for passes leading directly from the sea to this plateau. A route by the Skeena river was not then thought of. [*Vide* page 133, S. Fleming's report of 1877.]

ascended the stream issuing from Lake Killoosah, and saw a portion o the fine Valley of the Kitimat. This has not been referred to in Mr. Cambie's Report.—[At page 38, Appendix C, Report 1878.] Neither has any allusion to the Kitimat been made by Mr. Fleming's engineers, until last March, when Mr. Fleming received the following letter from the writer :—

"OTTAWA, 9th March, 1880.

" *Sandford Fleming, Esq.,*
  " *Engineer-in-Chief, Canadian Pacific Railway.*

" SIR,—Having in view Mr. Keefer's recent survey from the head of
" Wark Canal, up the Skeena, through the Cascade Mountains, with the
" object of making rail communication between the Forks of Skeena and
" Port Simpson, it has occurred to me to make the following suggestions;
" which, although rather late in the season, may prove interesting :—

" An inspection of Mr. Keefer's plan shows, as indeed might have
" been expected, more than fifty miles of extremely difficult location,
" through the very core of the coast range, which added to the distance
" from the Head of Wark Canal to Port Simpson, aggregates at least
" one hundred miles of the most expensive railway work, between the
" Kitsumkallum River and the suggested terminal point, Port Simpson."

" Now, I think it is possible to avoid this difficulty, simply by
" diverging from some point on the Skeena below Kitsellasse Canon,
" southward towards Lake Killoosah, and thence following the wide,
" open valley of the Kitimat to the Head of Douglas Channel, where I
" have no doubt whatever that it is possible, at an expense very much
" less than the difference in cost of construction between the Lower
" Skeena and Port Simpson route, and that now suggested, to form a
" good terminal harbour."

" In 1874, I examined the Kitimat Valley, for the purpose of finding
" an outlet in that quarter from the interior plateau. I was unsuccess-
" ful, although I pointed out the favourable features of this valley. In
" my report (see your own report for 1877), I gave a description of the
" Douglas Channel, but by some mischance that portion was omitted."

" In view of this circumstance, I deem it not out of place to again
" bring before you the above facts, which, it must be confessed, were
" not very clearly put forward in my report of 1874."

" I would add that the valley of the Kitimat is one of the most
" extensive on the coast, and I am confident that the summit between
" the Kitimat and the Skeena does not exceed 1,000 feet above sea level.
" Moreover, by this route, the formidable ' Cascades ' will be avoided
" altogether, and the distance between the Forks of Skeena and the sea
" shortened at least 50 miles."

"At the north-east corner of the Inlet, there is a natural harbour two miles in length, perfectly sheltered, but shallow. This could be easily dredged, were the main volume of the Kitimat diverted to the west side of the canal. The head of the canal is, of course, only a roadstead, but I think there is tolerably fair anchorage, and the offing can be reached by a magnificent channel and Nepean Sound."

"For steamships this harbour is as easily accessible as any on the coast. It seems to me that a proper hydrographical survey should be made, as also a survey from the head of the Inlet to some point on the Skeena near Kitsumkallum River, and should this harbour question be solved successfully, this route may prove even shorter than any yet suggested."

"I am, sir, yours, etc.,
(Signed), C. HORETZKY.

Mr. Fleming acknowledged the receipt of the foregoing as follows :

"OTTAWA, 10th March, 1880.

"MY DEAR SIR,—I find, in looking over Mr. Keefer's Report, now in type, he refers to the suggestion you made yesterday. In the second last paragraph of his Report he mentions the Valley of the Lakels (Killoosah) as offering easy access to Gardner Inlet. Mr. Keefer informs me that Mr. Cambie went to the lake near the summit in the year, 1877, and looked down the Kitimat Valley. I think I remember he discussed the matter with me at the time, but, for some reason or other, it went no further. I have just seen Mr. Keefer and he confirms all you say about the character of the Valley.

"Yours, etc.,
(Signed), "S. FLEMING."

Mr. Keefer neither saw the valley of the Kitimat nor the Douglas Channel, but he confirms all the writer says about its character, and adds that there will be no difficulty in carrying a line by it to the head of Gardner Canal. He is mistaken ; it is a physical impossibility to carry a line from the Kitimat to the head of Gardner Canal, or, in other words, to reach the head of Gardner Canal from the head of the Kitimat would involve 90 miles of the heaviest work along the roughest Canal on the coast, an engineering feat no one would ever dream of attempting ; but he may have mistaken one inlet for the other.

It is certainly an odd circumstance that the finest valley, without exception, upon the British Columbian sea-board, piercing the "Cascades," has been overlooked without any assigned reason. There is not the shadow of a doubt as to the possibility of making an excellent terminal port at the head of the Kitimat or Douglas Inlet. I consider that there will be from 500 to 600 acres of water area available for shipping in the north-east bay when dredged out. Wharves can be constructed all round it, and it will then form a fine dock, perfectly sheltered from every wind. Outside several square miles of water area can be made available by stretching a floating breakwater across the Inlet wherever suitable, and in the event of the water being too deep for anchorage, anchor buoys can be disposed as desirable. A floating breakwater of boiler plate, and in sections of any desired length, drawing three feet of water, amply sufficient to form a perfect mill-pond to leeward in that sheltered inlet, will not, exclusive of moorings, cost more than $150,000.

There is no inlet on the coast which offers greater facilities for the construction of wharves. The adjoining ground is certainly more convenient for the business of a large city than that around Port Simpson, being perfectly level, with room for extension twenty miles back, if necessary.

Owing to the fact of the upper harbour of Kitimat being completely land locked, and also to the large volume of fresh water which the easternmost mouth of the river Kitimat pours into it, there are times during winter when ice forms. This, the Indians averred, was always carried away by the tide. However that may be, the diversion of the Kitimat to the west side of the inlet, and the dredging out of the shoals within, and at the narrow entrance of, the upper harbour, will certainly obviate any inconvenience which might arise from that cause.

As regards the climate at the head of the Douglas Channel, it may be said to differ but slightly, if at all, from that of Port Simpson. The Douglas Channel is straight and wide, its upper extremity within fifty miles from Whale and Wright Sounds, and being thus more subject to the atmospheric influences of the Pacific Ocean than the long, tortuous and dismal Fiords of "Gardner," "Dean" and "Bentinck," which pierce the very core of the coast range, must necessarily be under similar climatic conditions as the more northern Port Simpson.

From careful enquiry, in 1874, the writer finds on referring to his notes that the average snowfall in the valley of the Kitimat rarely, if ever, exceeds four feet.

The climate of this northern coast has been much decried and unfavourably compared with that of the southern portion of British

Columbia. It is certainly more humid, but undoubtedly less subject to fogs than the southern Georgian Sea, and the coast, if wet, is no worse in that respect than those of Nova Scotia, the upper Atlantic States, and Scotland.

The reader may, in this connection, derive some valuable information from Captain Brundige's weather tables, showing the climate of Port Simpson, at pages 163 to 167 of Mr. S. Fleming's last railway report.

In addition to the facilities afforded by the lower portion of the Valley of the Kitimat for the site of a large city, the harbour and its approaches are admirably situated for defence, and can, with the greatest ease, be made completely safe from foreign attack. Beacon Hill, named by the writer, 1,450 feet in height, from which a photographic sketch of the upper valley and harbour was taken, commands the latter as well as the magnificent channel to the southward, besides being able to sweep the upper portion of the valley, in the event of any attempt at hostilities from the Skeena quarter. The citadels of Quebec and Gibraltar sink into insignificance when compared with this commanding and impregnable position. So much cannot be said of Port Simpson, which would be almost within range of Alaskan batteries in the event of war, and is by no means so favourably situated for defence.

Illimitable water power is available for mills and factories throughout the entire length of the Kitimat Valley, the Kitimat River and its eastern tributary, the Lachaques, affording a constant supply. In fine, as remarked before, there is no locality upon the whole British Columbian Coast line, which combines so many natural advantages for the Western terminus of the Pacific Railway.

Clio Bay, a few miles below the Head of the Inlet, on the eastern shore, has already been alluded to. There is fair anchorage there.

It has been stated that this Inlet is very readily accessible from the offing by Nepean and Wright Sounds, and from Port Simpson. It is also accessible by the Ogden Channel, a passage nearly mid-way between Douglas Channel and Port Simpson.

At page 154, S. Fleming's Report, 1880, Captain J. C. Brundige' thus speaks of it :—" I consider there is not a better locality for ships "to make the land on the whole coast than here."

. Port Fleming, at the upper end of the Grenville Canal, adds another to the list of the havens of refuge favourably situated for vessels approaching Douglas Channel. *Vide* page 154, Mr. S. Fleming's Report, 1880.

Speaking of the approaches to the coast, either at Port Simpson or south of it, Captain Brundige says, at page 159, Report 1880 :—

" Ships coming from the south and west can make Cape St. James " in safety, just as ships make Cape Clear, on entering Bristol, or other " channels."

" As they sail up, they can enter Ogden, Eddy or Brown's Passages, " either of which is superior to San Juan."

By " San Juan" he evidently means the passage from Fuca Straits to Burrard Inlet.

As one possessed of a very fair knowledge of the British Columbian coast, and of other foreign coasts and harbours, the writer can readily corroborate all Captain Brundige says, and unhesitatingly affirm that the "Douglas" Channel, at the head of which is situated the proposed terminal harbour for an Imperial and Canadian trans-continental railway, is as safely and as easily accessible from the Pacific Ocean as many of the very best Pacific Coast harbours, and infinitely more easy of approach than the harbours of Burrard Inlet.

To reach the " Kitimat," either from the Nepean Sound or the " Ogden " Passage, the towage for sailing ships would not exceed 60 or 90 miles in either case.

To reach the harbour of English Bay (Burrard Inlet), the same towage is necessary, but the risks of navigation are greater.

An inspection of Captain Brundige's report shows that he examined nearly every place of importance in the vicinity of Port Simpson but the head of the Douglas Inlet, a circumstance which, taken in conjunction with the mutilation of the writer's report of 1874, Mr. Cambie's silence on that subject in his Report of 1877, and the ignoring by the Chief Engineer of the matter in recent reports, must appear singular.

It is, however, by no means improbable that the examination recommended by the writer in his letter of the 9th March last to the Chief Engineer may be, even at this late hour, in process ; if so, it is to be hoped that the person entrusted with this work may not be amenable to paltry considerations, and that he will report conscientiously.

The writer considered it his duty to address the Premier of Canada, briefly, upon the subject which has formed the gist of this paper. Accordingly, Sir John Macdonald was written to on the 12th inst. (May).

Numerous surveys and explorations have been carried out within the last decade throughout all parts of the North-West. The writer has, as must now be apparent, taken part in those very important operations which have cost such enormous sums to the Dominion of Canada. It appears that those examinations might have been made without the prodigious outlay involved, especially on the Pacific side; it is not, however, the writer's intention to discuss that matter.

During the season of 1879, a very large and expensive expedition went into the Peace River country. The party consisted of Mr. H. J. Cambie, Mr. MacLeod, the Rev. D. M. Gordon, and Dr. Dawson, of the Geological Survey of Canada. The programme of their proceedings was to enter from the Pacific side, descend the Peace River, cross the Pine River Pass, and examine the Peace River country. With a view of meeting the Rev. Mr. Gordon on the east side of the Rocky Mountains, another expeditionary party was dispatched from Winnipeg to meet him as he emerged from the fastnesses of the Athabasca region.

Their outfit from Fort St. James was, in Mr. Cambie's own words (*Vide* page 42, Report 1880) :—

" Our party, for the exploration of the Peace River country, then "consisted of six on the staff (Mr. Cambie was also accompanied by a " Secretary), 14 packers, besides two men and 5 Indians, 27 in all, and " our train consisted of 72 pack-mules, with 23 riding animals; a total " of 95 animals."

It would be very interesting to the public to know the result of this grand expedition, but, as space will not permit us to follow each individual member of this party in his peregrinations, it will suffice for present purposes, to know what the chief of the expedition did during the long summer of 1879.

With the exception of Doctor Dawson, who crossed the Rocky Mountains, by the Pine Pass, the whole party descended the Peace River from Fort McLeod to Dunvegan, by boat and raft.

Mr. Cambie thence rode on horseback to Smoky River, (45 miles), from Smoky River to Sturgeon Lake, (41 miles), from Sturgeon Lake to Little Smoky River, (30 miles), from Little Smoky River to Lesser Slave Lake, (28 miles), from Lesser Slave to Peace River, (55 miles), from the crossing of Peace River along the left bank to Dunvegan, (50 miles), from Dunvegan to Fort St. John, (120 miles), and from Fort St. John to Hudson Hope, (40 miles).

In all this distance but a small portion was seen of the route proposed by the writer in 1872, *i.e.*, in the distance, between Dunvegan and Lesser Slave Lake going east. Mr. MacLeod performed the greater part of the examination.

From Hudson Hope, Mr. Cambie travelled to Moberly Lake, and thence to the Pine River (26 miles), up the Pine River (previously examined and favourably reported upon by Mr. Hunter) (35 miles), and from the Pine Pass Summit to Stewart Lake, (108 miles). Much of this distance was over ground travelled by the writer in 1872, and the result of this expedition has been (apart from the valuable information obtained by Doctor Dawson, and the minute inspection of the engineering features of the country south of the Peace River, by Mr. MacLeod), but to confirm the writer's views expressed in 1872.

In verification of this statement, I shall here quote some extracts from a Memorandum by the writer to the Minister of Railways, dated, " 20th January, 1879," at the request of the latter.

" In point of fact, the Peace River Pass is not so formidable as that, " nor in any portion does it bear any resemblance to a canôn, excepting " between the head and foot of the Rocky Mountain canôn, or portage, " which is entirely beyond, and *east* of the main range, and is on a very " much reduced scale, as compared with the steep, rocky slopes of the " main range."

" In reality, the passage of the Peace River, through the Rocky " Mountains, is an easier problem to solve than the continuation of a line " immediately to the eastward, in its low trough 700 feet beneath the " plateau, or in close proximity to the river along the adjacent heights, " the last alternative, all but impracticable, on the line indicated in Mr. " Fleming's Report."

" Indeed, were it possible to carry that line at the high level of the " plateau (1,700 or 1,800 feet above sea), there might be some reason " for taking advantage of the Peace River Pass, were it advantageously " situated with regard to western termini ; but, as any line through " that pass must either descend to the low level of the Peace " River, when *east* of the Rocky Mountains, and maintain that " level as far as the Smoky River, *or must diverge from the eastern* " *portal of the Pass southward*, in order to avoid the prodigious valleys of " the Whitefish, Pine, Mud, Échafaud and other streams, the objection " to the Peace River Pass may be readily understood."

"The line in discussion is traced across the portage, from the head
" of the Rocky Mountain Canôn, and easterly along the Peace River, to
" the Forks of the Smoky River."

" It is impossible to carry a line as projected across the portage
" without miles of tunnelling, for the simple reason that the lowest part
" of the ridge across which the portage trail is made, is 1,000 feet higher
" than the water level at Hudson's Hope, while to follow the semi-cir-
" cular course of the canôn (25 miles in length), will, in all probability,
" entail heavy works in rock, however, I do not question the practicabi-
" lity of the latter alternative along the right bank, but I am of the
" opinion, that from Hudson's Hope to the Smoky River, a line following
" the low level of the Peace River, will be objectionable, in consequence
" of unavoidably bad alignment, its length and heavy works occasionally,
" with frequency of bridge structures across the mouths of the southern
" tributaries and numerous dry ravines, not to mention the difficulty of
" gaining the high level again beyond the Smoky River."

" In all the distance from Hudson's Hope to the Smoky River, the
" Peace River flows at the bottom of a trough, 600 to 800 feet, beneath
" the surrounding plateau. Alluvial flats, terraces of varying altitudes
" alternating with rocky exposures, clay and gravel slides occupy the
" slopes of this trough, which are intersected by numerous rivers, count-
" less creeks, and dry coulees, all of which debouch on the main stream
" through immense gorges, showing vertical sections as deep as that
" of the Peace River Valley itself."

" The immense ravines and river valleys above mentioned, place
" road construction along the heights in proximity to the river *entirely*
" *out of the question.*"

" For the above reasons, I have always favoured a line from the
" Pine Pass, parallel to the Peace River, but thirty or more miles to the
" southward, where the valleys of the rivers are of decreased depth, and
" where they might be crossed to better advantage."

" Mr. Hunter's exploration of 1877 has, so far, proved the sound-
" ness of my views and has even exceeded my expectations."

Messrs. Cambie and MacLeod have amply verified the views expressed in the above extracts (see Mr. Fleming's Report, 1880, pages 44, 45, 56) but it is difficult to understand why such an extravagantly equipped party was needed, especially with the whole summer before them.

The writer's party of 1872 was but a toy affair compared with the expedition of 1879, which, in addition, was met from Winnipeg by another outfit of no mean proportions.

In September, 1872, the writer and his associate, Mr. Macoun, left

Edmonton with ten horses and two men *en route* for MacLeod Lake via the Peace River. Nothing was then known of the country excepting by the Hudson Bay people, but the little expedition made its way, and passed through the Peace River Pass at the most inclement season of the year, reaching MacLeod Lake in November. The whole cost of that examination did not exceed $1,000, and the results were highly important.

NOTE.—Cost of Horetzki's Expedition from Edmonton to McLeod Lake :—

Hire of 10 horses from the Hudson's Bay Company .. $450
" " 2 men, Edmonton to Rocky Mountain Canôn 100
" " 4 Indians and 1 half-caste, Rocky Mountain Canôn to McLeod Lake 207
" " Boat, (cash left with Hudson's Bay Company to reimburse owners) 20
Provisions and sundries 200
$977

Mr. Cambie had also a Secretary at a high salary, besides a Majordomo at $75 per month. The duties of this person were to talk to the Indians, to see to the putting up of the tents, etc., etc. Four or five of Mr. Cambie's men were sent home from the east side *via* U. P. and C. P. Railroads and steamer to Victoria, at a very great cost to the Government.

Let there be no misunderstanding ; Messrs. Dawson and MacLeod did excellent work, the former, by examining the country as to soil, geology and climate ; the latter, in surveying minutely the engineering features of the line proposed in 1872.

Mr. Cambie takes exception to Professor Selwyn's description of the country about Hudson's Hope.

The latter says at page 62, Geological Report of 1875 :—

"*11th September.*—A little white frost ; thermometer : 32° at 6 a.m.
" Charlette lent us two horses, and, at 7.30 a. m., Mr. Webster and I
" started for Moberly's Lake. The trail runs two or three miles up the
" river, and then turns to the left and ascends by several steps or
" benches to the plateau ; an undulating country of alternating low,
" sandy or gravelly ridges, covered with forest of small pine, and swampy
' depressions, with spruce and tamarac and well-grassed flats, thinly-

" wooded with aspen, alder and willow. In places the woods were all
" burnt, and in these *brulés* we lost a good deal of time searching for the
" trail. At 6 p. m., however, we reached the top of a hill, from which a
" small piece of the Lake was visible, about three miles distant down a
" narrow valley. We camped here, an operation which consisted in
" lighting a fire, putting up a few boughs for a break-wind, and eating
" our supper of bread and dried moose meat. Starting at 7 a. m. on
" the following morning, we reached a rocky hill immediately above the
" south-west end of the lake at 9.30. The trail wound round the flank
" of it and descended towards the lake shore. As there was nothing to
" be gained by following it further we dismounted, and, leaving our
" horses on the trail, climbed to the summit, where an extensive view of
" the surrounding country was obtained and a series of bearings taken.
" The barometer reading was 26.59, indicating about 2,000 feet above
" Hudson's Hope, and only a little less elevated than Table Mountain
" on Pine River, which I think I recognised, bearing 97°. The strata
" here are quite similar to those of Table Mountain—horizontal, thick-
" bedded, reddish-brown and grey sandstones—but no fossils were seen
" in them. The hills around the lake, especially the lower slopes and
" the intervening valleys, are richly grassed. Pea-vine, *Astragalus* and
" various nutritious grasses standing above one's knees on horseback.
" *There are large areas of open prairie land*, and more which is only
" wooded with willow, aspen and alder copices. On the higher slopes
" pine prevails, and, in the low grounds, spruce, tamarac and poplar.
" A purple-red *Epilobium* is very abundant, also service-berry, 'poire'
" and a species of *Viburnum*—high-bush cranberry. I never saw the
" berries on the latter so fine or so abundant. On some of the open,
" sandy ridges, blueberries and cranberries were also plentiful. Charlette
" tells me that the snow fall is comparatively light, and that horses do
" well through winter amongst these hills. *I consider it a region far
" fitter for settlement than much of the Saskatchewan country.* We are now
" in the middle of September, the thermometer has only once reached
" 32°, and potatoe tops at Hudson Hope are still green."

At page 51, Pacific Railway Report, 1880, Mr. Cambie thus describes the same locality:—

" We reached Hudson's Hope September 15th, and tried to obtain
" a guide to take us to Pine River, but failed, as the Indians were
" all absent; accordingly we left next morning and followed a
" hunting trail to Moberly's Lake. The trail ascends from Peace River
" by a series of benches, and at one and a-half miles reaches the plateau,
" which is here about 2,000 feet above sea level, and continues at the
" same elevation to the fifth mile; it then passes over a ridge 900 feet
" above the plateau and along a steep hill side to the south-western end
" of Moberly's Lake, at an estimated elevation of 2,050 feet above sea
" level."

"According to the best sources of information at my disposal, (Mr. Selwyn's Report) "Moberly's Lake should have been situated two-thirds of the way across from the Peace to Pine River, and in a country fitted for settlement, though somewhat hilly and with large areas of prairie land."

"Great was my surprise, therefore, to find myself only nine miles from Hudson's Hope, and hemmed in by hills, rising from 3,000 to 4,500 feet above sea level, the only level land visible, being in the valley of Moberly's River, which empties into the lake from the west; and further, that between me and Pine River lay a range of mountains at least sixteen miles broad, rendered almost impassible by fallen timber, the only prairies being on the slopes of steep hills facing the south."

"There was no possibility of retreat; the party on Pine River was waiting for us; and, having only a limited supply of provisions, delay might prove disastrous to both parties."

"Fortunately, I was able to reinforce my little band by engaging the services of an Irishman named Armstrong, whom we found building a shanty for himself in order to hunt during the winter; he had spent part of the summer at the lake, hunting, prospecting for gold, and catching fish for the support of a number of sleigh dogs belonging to the Hudson Bay Company."

"White fish were then, [September 17th,] very abundant, and he gave us all we could carry. They varied from 4 to 6 lbs. in weight, were very fat and seemed to me quite equal to the far-famed white fish of Lake Huron."

"We followed the valley of Moberly's River, south-westwards, for eight miles and then turned southwards up a small tributary. After four days, during which we had chopped our way through fallen timber from day-light to dark, I found myself in a small basin with hills rising steeply 1,000 to 1,200 feet on both sides and in front, and these, where not actually precipitous, were so strewn with fallen timber of large size, that it seemed a hopeless task to attempt to cut our way through without help. I therefore sent two men ahead to find Mr. Major and get some of his party to come to our assistance, while I remained behind to take care of the mules, assisted by Armstrong, who had cut his foot with an axe."

"My messengers returned three days afterwards with six men, and on September 24th, we reached Pine River and joined the main party."

"I estimated that we were only 17 miles from Moberly's Lake, but had travelled nearly 30, and in the last four miles had passed over a mountain 4,200 feet above sea-level. We were also 21 miles west of the point where I expected to find myself."

" In the first five miles from Hudson Hope, we had crossed two
" small tamarac swamps and some stretches of light, sandy soil, with a
" small growth of poplar and spruce."

" We had again met with some level land in the Valley of Moberly's
" Lake, which for nine miles above the lake averages nearly half a mile
" in width in the bottom. Some portions of this are gravelly and barren
" and others fertile, with a few small prairies producing rich grass
" There are also some fine prairies at the lake, on slopes facing the south.

Mr. Macoun also remarks of this locality, page 152 of Geological Report, 1875 :—

" The following extract from my journal, written on the spot, will
" give a truthful picture of Hudson Hope, as I found it on the 22nd
" of July, 1875 :—I have been extremely surprised at the rankness of the
" vegetation around here, although there is very little rain at this season
" and has been little all spring. Wild peas and vetches grow to an
" amazing height in the poplar woods, and form almost impenetrable
" thickets in many places. Vetches, roses, willow-herb and grasses of
" the genera *Poa*, *Triticum* and *Bromus* fill the woods and cover the burnt
" ground, and surprise Canadians by their rankness and almost tropical
" luxuriance. Charlette, who is in charge of this post, has two small
" gardens, in which he has growing, potatoes, onions, turnips, beets,
" carrots, cabbage, and various other vegetables. Yesterday we had new
" potatoes for dinner, of a very fair size, which were planted on April
" 28th. Numbers of the onions were one and a-half inches across, raised
" from seed imported from England, and sown about the 1st of May.
" Growth is extremely rapid, owing partly to the length of day and
" cloudless skies supplemented by heavy dews, and possibly also in part
" to the great range of temperature during the 24 hours, from about 45°
" at sunrise to 80° Fahr. at noon. Sometimes the range is even more,
" but the above may be taken as the average. The rankness of the
" vegetation on the west shore of Lake Superior has frequently been
" alluded to, and may be caused by the somewhat similar great range in
" the temperature there. Can it be that all the rank vegetation observed
" around Lake Superior, in the Rocky Mountains and here, is connected
" with the sinking of the temperature during the night, and increased
" activity given to the vegetation during the day on this account ? We
" have warm sultry days, and cool pleasant nights, with constant regu-
" larity, and we are told that this is the usual summer weather. The
" left bank of the river is much drier than the right, and, as a consequence
" of this, growth on it is much further advanced. The frost of the 28th
" of June, however, was more severe on the left bank than on the right.
" Charlette informed me that in 1874 there was no frost from the 1st of
" May to the 15th of September. In 1875, sowing commenced the last
" week in April, and the first frost came on the 8th of September."

From the testimony cited in the foregoing pages, it must be manifest that the final choice of the southern Burrard Inlet route may involve consequences of dire import to the Dominion of Canada.

It will, doubtless, be said in its justification, that the bulk of the population, indeed, the entire population of the mainland, is centred along the Frazer River and its tributaries in the south. True. But the personal interests of 3,000 whites should not be allowed to weigh against the general well being of the rest of British North America.

A line terminating to all intents and purposes, upon United States soil, (at Holme's Harbour), and running for six hundred miles through an irreclaimable wilderness is not what Canadians bargained for, it is not what British Imperial interests demand, and Canadians, if loyal to themselves and to Great Britain, will see to it.

As for the plea that the construction of 125 miles of railway between Yale and Kamloops (to cost $12,000,000, and as much more as the contractors may choose), is only intended for local purposes, that is to say, for the purpose of serving the traffic of the interior plateau, which has never yet exported a bag of flour, of which the sole agricultural exports for 1878 were cranberries, valued in the aggregate at $462, then nothing more need be said.

The canned and pickled salmon of the Lower Frazer, the lumber trade of Burrard Inlet, the coal of Nanaimo, require no railroads. They have the finest waterways in the world to serve their purpose.

The following facts are significant:

In his Report of 1879, Mr. Fleming penned the following paragraph, which he has quoted at page 4 of his latest report:—

"It cannot be said that the selection of Burrard Inlet as a terminus,
" has given general satisfaction in British Columbia. On the contrary,
" a claim has been advanced in that Province that another route and
" terminus are preferable. It is, therefore, to be considered if additional
" explorations should be made and more complete information obtained
" with regard to the northern country, so that it may be definitely deter-
" mined if a route more desirable can be found. Accordingly, I suggest
" that the unexplored region, lying between Fort Connelly and Fort
" McLeod, in British Columbia, and those large tracts of vacant terri-
" tory east of the Rocky Mountains, in the latitude of Peace River, which

" have never yet been traversed by scientific travellers, be explored, and
" accurate data obtained respecting the feasibility of a railway through
" that region to the Pacific Coast."

He then proceeds to say in the Report of 1880 :—

" The Burrard Inlet route was known to be marked by many diffi-
" culties, and to involve an enormous outlay, but with all the disadvan-
" tages which it presents, I considered that it was entitled to the pre-
" ference.

" For six consecutive years, and at an exceptionally great cost,
" unremitting and systematic efforts had been made without success to
" find a better and less expensive line. Indeed, there seemed no alter-
" native but the adoption of that route, unless further examination of
" the northern country made it apparent that a better and more eligible
" location could be found under conditions so favourable that it would
" command ready acceptance.

" Owing, in some degree, to the fact that the northern districts of
" British Columbia are remote from the areas of population, a northern
" route obtained but little attention during the early stages of the survey.
" It was only when it was found that no line could be secured in the
" more southern latitude, except at great outlay, that a northern route
" came prominently into notice, and that more extended examinations
" became desirable."

" It was a serious responsibility for any engineer to assume to
" recommend that construction should be commenced on the line to
" Burrard Inlet, without first having exhausted all the sources of enquiry
" open to us. I felt that we should clearly and unmistakeably under-
" stand the capabilities and possibilities of the northern region, that we
" should obtain data, to determine if a railway line could be obtained
" through it, that we should know the character of the route, and that we
" should possess full information with regard to the climate, soil and
" capability for settlement, before the Government became irrevocably
" committed to the large expenditure attendant upon the adoption of any
" route."

" *It is easy to be understood that, if, subsequent to the construction of
" the railway on the southern route, it was discovered that a northern line
" could have been undertaken at a greatly reduced cost, through a country, in
" respect of soil and climate, suitable for prosperous settlement, a gross and
" irremediable error would have been committed, possibly ever to be deplored.*"*

" Additional northern explorations, therefore, seemed to be advisable
" whatever the result obtained. Under any circumstances, it was evident
" that the information gained, even if of negative value, would be
" important in adding to our positive knowledge of the territory."

---

* NOTE.—The reader's attention is particularly requested to this paragraph.

"In April last, I was notified that the Government had decided,
" previous to the determination of any route, to make additional examin-
" ation of the northern passes and of the country which flanks both sides
" of the mountains."

" These examinations it was proposed so to carry on that the
" information would be systematically and rapidly gained, that it could
" at once be acted on and the choice of the location, and the commence-
" ment of construction, no longer delayed."

" The extent of territory embraced was the country between the
" longtitude of Edmonton, east of the Rocky Mountains, and Port Simp-
" son, on the Pacific. Port Simpson had already been reported to be an
" excellent harbour. It was known that a deep-water arm of the sea,
" named Wark Inlet, some 35 miles in length, extended to the east of
" Port Simpson, in the direction of the River Skeena ; Wark Inlet being
" separated from the Skeena by a narrow isthmus of no great elevation.'

" The objects of the examination were to discover the most favour-
" able route from the coast to the Peace River District, on the eastern
" side of the mountains, and thence to the line already located near
" Edmonton ; to gain full information with regard to Port Simpson, to
" verify the reports as to Wark Canal being navigable by ocean-sailing
" ships, to ascertain how far the country lying between the head of that
" sheet of water and the River Skeena, and the Valley of the Skeena itself,
" were suitable for a railway line, and to obtain such definite informa-
" tion respecting the nature of a portion of the line accessible to steam-
" ers from the ocean, as would admit of a contract for construction being
" at once let, in the event of a northern route being chosen."

" The examination really involved the determination of the problem
" whether the choice of the Burrard Inlet route should be sustained or
" abandoned, and if construction should be immediately commenced on
" the northern or on the southern line."

" The service was, consequently, one of importance. The instruc-
" tions to the officers selected, together with their Reports, are given in
" full in the Appendix. As time was an element in the problem, it was
" arranged that the examinations should be energetically carried out,
" and that so soon as the information was obtained, a synopsis of it
" should be sent by telegraph from Edmonton to Ottawa. Before the
" end of September the information was received and laid before the
" Government."

The writer was one of the " officers selected " to carry out these examinations. His duty was the most arduous, and involved the exploration of a route through three distinct mountain ranges, across a hitherto unknown portion of the country, entirely on foot, and by canoe where possible. *The section he had to examine was in fact the key to the whole question of a route from Port Simpson to the eastern prairie region, as his objective point was the Peace River Pass,* the Pine River route having been counted out. (*Vide* pages 9 and 10 Pacific Railway Report of 1878, whereon Mr. Fleming says) :—

"I do not attach the same importance to the 'Pine River Pass' as
"Mr. Smith. * * * Although favourably situated for a line to a
"Northern Terminus, its importance is not enhanced by the fact that
"a still lower pass—Peace River—exists, only a few miles further north.
"I have accordingly projected a northern line of railway through Peace
"River Pass, which I consider preferable."*

On the 25th July, 1879, the writer indited the following letter to Mr. H. J. Cambie :—

"HAZELTON, FORKS OF SKEENA, 25th July, 1879.
"*H. J. Cambie, Esq.,*
"*Stewart's Lake, or elsewhere.*

"DEAR SIR,—I have just returned from a preliminary reconnais-
"ance, *via* the Atnah Pass, Bear Lake, Driftwood River, Lake Tatla,
"and the 'Frying Pan Pass.'

"I have reached the following conclusions : that, if the Babine
"River prove as favourable as I suppose it will, a direct and generally
"easy line will be found from its upper portion, easterly through a fairly
"level country, to the Kotsine River, thence across the Driftwood River
"Valley, and from Tatla Lake to the Omenica, by one of several low
"passes, either touching the Omenica-Sitlica, or through the present
"pass, used by miners, which I believe available for a railway.

"I have reason to think that the lower Omenica will prove favour-
"able. My preconceived ideas regarding this northern country and the
"central range through which the Frying Pan Pass leads, have been
"considerably modified since my recent journey, and I believe that the
"profile of this proposed line will prove better than any yet found, ex-
"cepting that of the Wotsonqua valley.

"There is quite a break in the central range in latitude 55° 40', and
"through this gap flows the Kotsine River, which I believe will afford a
"good pass (probably not higher than 3,000 feet above the sea) from a
"level valley which I have seen from mountain heights to the north-
"ward, and which communicates with the upper portion of the Babine
"River.

"Eastward from the Driftwood River I anticipate no difficulty in
"getting over into the Omenica, as I have already stated.

---
* NOTE.—The writer always attached more importance to the Pine River Pass than to that of the Peace River, and Mr. Cambie has been obliged to admit that the former is the better of the two.

" The only objection I yet see to this route is the circuitous course
" of the Babine River, but that drawback would be common to all other
" lines via the southern end of Lake Talta.

" The mouth of the Babine River is at least forty miles due north
" from this point, and Fort Connolly is in a higher latitude than is
" shown on the map. On my return I have carefully examined the
" Susqua Valley, and have taken more precise heights of its summit,
" which I now place at 1,400 feet above Lake Babine, and 3,100 feet
" above Hazelton.   *   *   *   *   *   I have, however, such
" good hopes of getting up the Babine River, between the mouth of which
" and the Lake outlet, the ascent will probably not exceed 1,200 feet,
" that further reference to the Susqua valley route need scarcely now be
" made.

" I expect, D.V., to reach " Hogem" on the Omenica, about the
" end of September.

" I am, dear sir, yours,

(Signed) " C. HORETZKY."

The substance of this letter was forthwith transmitted to Mr.
Fleming at Ottawa, and must have reached him some time in August.

Meanwhile, the Peace River party descended the Peace River, and
telegraphed the results of their examination as follows :—

" TELEGRAPHIC REPORT ON EXPLORATIONS FROM FORT
" SIMPSON, ON THE PACIFIC COAST, TO EDMONTON, VIA
" THE PEACE RIVER VALLEY AND PINE RIVER PASS,
" BY MESSRS. CAMBIE, MacLEOD, DAWSON AND GORDON.

" To Sandford Fleming,
" Ottawa.

" From Hay Lake, 24th September, 1879.
" (Near Edmonton.)

" Arrived last night. Reached Dunvegan 1st August; left there
" 2nd September. Party spent month in exploring country. Tupper
" left Edmonton 8th August. I came by Slave Lake and Athabasca
" Landing. Country from ten miles south of landing to Edmonton
" excellent on both sides of road, improving towards Edmonton ; chiefly
" prairie with aspen copse and occasional pine and spruce. Distances :
" Dunvegan to Smoky River Post, 50 miles ; Smoky River to Slave
" Lake Post, 62 ; Slave Lake, 70 ; east end Slave Lake to Athabasca
" Landing, by river, about 120 ; Landing to Edmonton, 96. My letters
" all forwarded, by mistake, to Tupper's care ; none received ; anxious
" to return homewards, without further special examinations. Country

" around here appears superior to Peace River country for raising grain.
" Before leaving Dunvegan all agreed on the following telegram :—

" Red line, letter A, to Slave Lake, direct and generally easy.
" Pine River, 500 feet wide; height of bridge, 70 feet. Gradients
" leaving the river, 1 foot per 100. Summit eastward, 900 feet lower
" than Hunter's, and 15 miles further north. Mud River, 400 feet
" wide; height of bridge, 60 feet. Gradients on west side, very easy ;
" on east side, 1 per 100.

" Échafaud River, 300 feet wide ; bridge, 60 feet high. Gradients,
" moderate ; work occasionally heavy three miles on each side of bridge.

" River Brulé, 50 feet wide ; bridge 70 feet high. Valley, narrow ;
" gradients, easy.

" Smoky River, 750 feet wide ; bridge 100 feet high. Valley about
" 500 feet deep at crossing ; gradients, slightly exceeding 1 per 100.
" Works very heavy for three miles on each side.

" Goose River, 400 feet wide ; valley, 200 feet deep ; bridge, 50 feet
" high. Gradients on each side easy.

" Whole country, from Pine River to Slave Lake, with these excep-
" tions, very favourable.

" Pine River and Slave Lake appear to be approximately correct on
" plan of 1876, but Smoky River and Dunvegan are placed about 50
" miles too far west."

" Blue line, letter A, to Southesk, examined to suitable crossing of
" Smoky River, latitude 55° 10', longtitude 118° 40' on the map of 1876.
" Blue and red lines, common to River Dechafaud, 50 miles east of Pine
" River ; thence to Smoky River, generally very easy. except about four
" miles following up the south Bank of the Échafaud River, where
" work would be heavy. No important streams crossed between Pine
" River and Smoky River. Approach to Smoky River by valley of large
" stream on each side ; bridge, 500 feet long, 60 feet high. Cannot
" report on remainder of line, not having heard from Tupper. Have
" ascertained that he was still at Edmonton, on 2nd August."

" Line from Fort St. James to Fort McLeod, undulating, but pre-
" senting no great difficulties as far as Long Lake, thence to McLeod's,
" following valley of Long Lake River. Gradients, long, 1 per 100 ;
" works very heavy, chiefly in gravel and stony ridges. A moderately
" direct line can be had from Fort Frazer to Fort St. James."

" Assuming direct line from Southesk to crossing of Smoky River,
" the route by letter A to Fort Frazer would be about 55 miles longer
" than the located route."

" Country pretty thoroughly explored as to general features from
" Pine River to Lesser Slave Lake, between 55th and 56th parallels of

" latitude, also for 70 miles north of Dunvegan. Elevation of plateau,
" generally below 2,000 feet. West of Smoky River, soil almost every-
" where very fertile. Extensive areas of prairie and lightly wooded
" country south of Peace River to near 55th parallel, also 50 miles or
" more northward. East of Smoky River, also fertile, but with very
" little prairie, and with many swamps and beaver dams, which could be
" generally easily drained."

" From information received, summer frosts occur occasionally in
" June, very seldom in July. We have experienced several in August,
" both in the valley and on the plateau. Wheat thrives and ripens at
" Hudson's Hope, Dunvegan and Lesser Slave Lake, the latter locality
" being on the level of the plateau."

" The party regard this statement as approximately accurate, but
" regret that pressure of time prevents the preparation of fuller and more
" satisfactory details as a special opportunity has arisen for forwarding
" this message from Dunvegan to Edmonton."

" At date of this Memorandum, 2nd September, all members of the
" party were well. Cambie returns with pack train by Pine River.
" MacLeod and Dawson continue explorations eastward and south-east-
" ward."

     (Signed) " H. J. CAMBIE,
          " HENRY A. F. MacLEOD,
          " GEORGE M. DAWSON,
          " DANIEL M. GORDON,"

The writer desires to point out that the above telegram proves beyond a doubt, the soundness of the views expressed in 1872, officially, and in " Canada on the Pacific," see map therein, where *" red line, letter A to Slave Lake,"* referred to in the above telegram, corresponds exactly with that shown on the map, and described to the Minister of Railways in the Memorandum dated, 20th January, 1879.

A few days after the receipt of this despatch, Mr. Fleming addressed the Minister of Railways, as follows :—

    " CANADA PACIFIC RAILWAY.
     " OFFICE OF THE ENGINEER-IN-CHIEF,
       " OTTAWA, 30th September, 1879.

" SIR,—I have the honor to report progress on some of the explora-
" tions authorized by you last spring, in connection with the Canadian
" Pacific Railway.

" Before deciding on the route through British Columbia, it was

" deemed advisable to gain additional information regarding the north-
" ern portions of that Province, as well as the territory extending east of
" the Rocky Mountains and lying between the latitude of Peace River
" and Edmonton.

" I have received despatches from several of the parties who were
" sent in May last, under special instructions, to explore in these
" regions, and who were directed to examine the harbours on the northern
" coast of British Columbia and the approaches thereto by sea.

" At the date of last returns, these examinations were by no means
" complete, but considerable progress had been made, and the informa-
" tion so far obtained is of importance.

" The country south of Peace River, hitherto unexplored, between
" the Rocky Mountains and Lesser Slave Lake, has been traversed in
" various directions as far south as the 55th parallel of latitude. The
" general character of the district for railway construction has been
" ascertained, and the fertile nature of the soil has been found to extend
" over a wider area than had been previously known.

" I have not heard from all the parties ; I cannot, therefore, refer
" to the explorations which by this time may have been made to the east
" of the mountains between the 55th parallel and Edmonton.

" Nor can I allude to the progress of explorations on the western
" side of the mountains between Fort MacLeod and Fort Connelly,
" embracing the basin of the Nation River.

" Although the examinations are incomplete and the returns partial,
" nearly all doubts are now removed as to the possibility of getting a
" practical railway line from the neighbourhood of Edmonton, by way
" of Peace River, and the valley of the River Skeena, to Port Simpson.
" The coast examinations, too, go to show that at Port Simpson a har-
" bour exists, which is probably unrivalled in British Columbia."

" The question of distance is an important one. The more northern
" route has not been instrumentally surveyed, and, consequently, the
" distance to Port Simpson cannot yet be accurately stated. A rough
" estimate, however, indicates that the line referred to, from Edmonton
" as a common point *via* the Peace River country, will probably be found
" 100 miles shorter to Port Simpson than to Esquimalt.

" The engineering character and the cost would, at the same time,
" I feel certain, be greatly in favour of the line terminating at Port
" Simpson."

" In comparing the line to Port Simpson, to which I have alluded,
" with the one *via* the Yellow Head Pass, to Burrard Inlet, the latter
" appears to be from 160 to 190 miles shorter, *but one of the advan-*

"tages which may be claimed for the more northern route is, that it would pass through and accommodate the Peace River country.* The line by the Yellow Head Pass could, with a branch, meet the same object, but to serve the Peace River district equally with the main line passing through it, the branch would be fully 300 miles in length. If we assume that this extra distance be added to the line to Burrard Inlet, we shall have both lines placed nearly on an equal footing, in point of mileage."

"The gradients on the route to Port Simpson would compare favourably with those on the line to Burrard Inlet, and I have reason to think that the total cost would be considerably less than the latter when the branch is taken into consideration. My previous reports give full explanations regarding the favorable geographical position of Port Simpson in relation to the Asiatic continent."†

"There can be no doubt that the examinations made this year, of which partial returns only have as yet been received, go to show that the northern route possesses advantages greater than previously known. From what has been brought to light, I would consider it unwise, at this stage, to adopt, and begin construction, on either the Burrard Inlet or Bute Inlet routes."‡

"While I would deem it prudent to defer a final decision with regard to the adoption of any route, until we receive more definite information regarding some portions of the country now under examination, I have no hesitation in saying, that, considered apart from the question of climate, *the route to Port Simpson presents itself with so many advantages that, to my mind, it opens up an excellent prospect of securing the most eligible route from the prairie region to the Pacific Coast.*"

"I have mentioned that the returns from our exploring parties are incomplete. From what I have learned, however, I am sanguine enough to think that, before the close of the season, we may have data to show that a line may be secured from the Peace River District to Port Simpson, considerably shorter than the line which I have above referred to. Should this view be realized, the comparison of routes will be still more in favour of the one terminating at Port Simpson.§

"With regard to the question of climate, I have, in previous reports, alluded to this subject. I now beg to refer to extracts from the letters of Capt. Brundige, a nautical gentleman specially detailed to

---

*NOTE.—Precisely what the writer has urged during the last eight years.

†NOTE.—Comparisons of cost will be found immensely advantageous, as regards the Northern line, and *without* the branch spoken of, while the Geographical position of the Kitimat is as favourable as that of Port Simpson.

‡NOTE.—With all deference to Mr. Fleming, the advantages of the Northern route *were* known before.

§NOTE.—Mr. Fleming refers to the substance of the writer's letter to Mr. H. J. Cambie, of date, 26th July, 1879, relating to the Kotsine Pass route.

"make full examinations and enquiries respecting the coast, harbours
"and approaches. I also append some extracts from Mr. D. M.
"Gordon's letters; that gentleman speaks for himself, and Messrs.
"Cambie and MacLeod, in regard to the explorations they have been
"engaged in in Northern British Columbia.

"From these it would seem that, while the interior of the country
"is free from an excess of moisture, the rainfall on the coast is great, and
"the climate there may compare generally with the west coast of Scot-
"land and with parts of Nova Scotia. From these extracts it will also
"be learned that Port Simpson is a capacious and safe harbour, and that
"it is perfectly easy of access to ocean steamers or sailing ships, night
"or day, and at all conditions of the tide.

"It is obvious that Port Simpson is a place which possesses excep-
"tional natural advantages, and in the event of a northern route for the
"railway being chosen, it would undoubtedly become a place of great
"importance. I would, therefore, suggest that no time be lost in taking
"steps to have the land in the neighborhood reserved.

"I have the honor to be,
"&c., &c., &c.,
(Signed) "SANDFORD FLEMING,
"Engineer-in-Chief.

The Hon. Sir CHARLES TUPPER, K.C.M.G.,
 &c., &c., &c.,
"Minister of Railways and Canals."

The foregoing letter shows conclusively that, even with the indefinite information received up to its date, Mr. Fleming felt that the northern route presented immense advantages—engineering and otherwise—over the Burrard Line. Port Simpson is the finest harbour on the whole coast, but to reach it, the formidable "Cascades," for a distance of 75 miles, must be passed.

Notwithstanding this great disadvantage, Mr. Fleming still saw the immense superiority of the northern route *via* Pine Pass, over the Burrard Line.

What, then, would he have thought of it, had he known that the sea could be reached without running the gauntlet of the coast range at all, simply by taking advantage of the Valley of the Kitimat? Without this knowledge, his letter to the Minister of Railways is unmistakeably in favour of deferring construction. And he had not then heard the writer's final report upon the missing link between Hazelton and the Peace River Pass.

With the imperfect data then in his possession, Mr. Fleming strongly, unhesitatingly, exhorts the Minister of Railways to defer construction on the southern line, but despite this professional advice proffered, it must be supposed, in good faith, an Order in Council was passed on the 4th October, ratifying the selection of the Frazer River line.

What is to be thought of this? And it would be interesting to know if Mr. Fleming had forgotten all about the Kitimat. The writer reported upon it in 1874, Mr. H. J. Cambie spoke of it *privately* in 1877, and had it again under his notice in 1879, yet the subject was completely ignored.

This matter is certainly worthy of full investigation for many reasons:—

The Kitimat, Pine Pass route is the easiest of all the lines hitherto examined in British Columbia. It presents fewer miles of heavy work than any other.

Between Rocky Mountain Summit and the Pacific its construction will cost (approximately) ten millions dollars less than the Burrard Yellow Head route. It passes through the dreaded coast range by a wide, open valley, the finest on the coast.

East of the Rocky Mountains it opens up a nearly continuous belt of agricultural and pastoral land, all the way from the Forks of Pine River to Manitoba.

Its general profile is the finest across the North American continent from ocean to ocean. Its highest summit is only 2,800 feet above the sea level.

On the Pacific slope it taps *the greatest connected region susceptible of cultivation in British Columbia* (*Vide* Dawson's report) which has a climate similar to that of Edmonton, where wheat attains perfection. (*Vide* Macoun's Report.)

At its western end there are unlimited facilities for the growth and extension of a large city.

Its terminus is only 4,000 miles from Yokahama, being 400 miles nearer to Japan and China than Burrard Inlet. That terminus is easier and safer of access than any proposed to the south, and Captain Brundidge expresses the opinion that the passages leading to it, and its approaches from the sea, are the best on the coast. (*Vide* page 154,

Railway Report 1880.) It is within ten or twelve hours steaming of Port Simpson, the best harbour on the British Columbian coast. It has, within easy reach, numerous havens of refuge. With the wind at west, south-west or south-east, sailing ships can reach the head of Douglas Channel, *via* Nepean Sound, *without* towage.

With a light-house on Cape St. James, and three other lights in the inner passages, the coast can be made on the darkest night with perfect safety.

In none of the numerous channels leading from the ocean to Douglas Inlet, are there any tide-rips or overfalls, the tide setting regularly along the coast, and rarely, if at all, exceeding a rate of three knots per hour.

Certain marine engineering works will be necessary to form a perfectly good harbour at Kitimat. Those have been referred to in the preceding pages.

It is clear that a northern route terminating either at Port Simpson or at the Kitimat, will be cheaper by millions of dollars than the Burrard line.

It is also evident that, to answer the purposes of a Colonization road, the northern line is infinitely preferable to the southern route, which must run for six hundred miles through an irreclaimable wilderness.

It is, or should be, intelligible to all, that, to carry a great colonization and imperial highway out of its proper course, upon the plea of serving the interests of 2,000 or 3,000 whites on the Frazer River, is absurd.

The writer feels that, strengthened as he is by the written testimony already cited, and backed by the evidence of the Chief Engineer himself, in his letter of the 30th September, 1879, addressed to Sir Charles Tupper, the ground he has taken in support of a northern route is impregnable.

In July, 1878, an Order in Council was passed, practically adopting the Burrard Inlet route. The late Premier had, acting upon the advice of the Chief Engineer of the Pacific Railway, authorized this action.

It has been shown, conclusively, that all reports upon the Kitimat Valley and route had been suppressed, and the inference is, that the matter had never been discussed between the Premier and his Engineer.

Indeed, the " Kitimat " had been systematically covered up, and hidden from everyone.

The probabilities are that, had Mr. Mackenzie been made aware of the existence of a fine valley through the coast range, as indicated, a thorough and exhaustive survey of the northern route would have been made before the adoption of a southern line. Mr. Mackenzie never had that information, consequently, he had no alternative but to follow his Engineer's suggestion, and adopt the Burrard line.

After the change of Administration in September, 1878, Mr. Fleming again urged the necessity for northern surveys, deploring in forcible language the serious consequences of a possible mistake in the choice of routes. As has been shown by the evidence given in the preceding pages, a cheaper and better route than that of Burrard Inlet was found.

If the reader will once more refer to the Chief Engineer's Report of the 8th April last, a strange and rather significant omission will be observed. The remarkably clear and very pronounced letter of 30th September, 1879, from Mr. Fleming to Sir Charles Tupper, urging the imprudence of adopting, or beginning construction on, the Burrard Inlet Line, in view of the examinations of 1879 by the Peace River party, is nowhere alluded to in that Report.

The importance of the letter in question being so great, and Mr. Fleming's opinion therein expressed so very decided, it is surprising that it should have been overlooked. Its omission from a report intended for the public is, under the circumstances, tantamount to an unequivocal withdrawal, and the public, having access to the report alone, must read the omission in such a light.

The question then arises : Had the Burrard Inlet line been decided upon beforehand, at all hazards, regardless of incalculable future injury to the Dominion ? Were British Imperial interests—an important factor in the railway scheme—to be sacrificed, by adopting a line terminating on United States soil, or, at the best, under the very guns of San Juan ?

Were the blunders of former Boundary Commissions to be supplemented by another, still further aggravated by the fact that its committal is actually taking place in full view of recent knowledge gained at great expense, *and in direct opposition to the Chief Engineer's vigorous protest of the 30th September last ?*

Were the explorations of 1879, then, a mere sham ? These explorations were solemnly, avowedly undertaken for the express purpose of averting a possible error, a calamity " ever after to be deplored " as Mr. Fleming gravely wrote. [See page 5, of Report.]

The results of these explorations are glaringly apparent, and point unmistakeably to a far better route for colonization purposes, and also one much easier of construction, and consequently less costly ; yet, notwithstanding, the Order in Council of July, 1878, was ratified, and a report framed in accordance.

The whole matter is certainly well worthy of a searching investigation, and in the meantime may afford the taxpayers ample food for reflection.

Within the past few weeks there have been rumours of a proposal, on the part of the Dominion Government, to hand over fifty millions acres of land in the North-West, to a company of English capitalists for the purpose of building the Pacific Railway. Recent movements of Ministers appear to confirm the truth of the report, and it is not unlikely, ere many weeks elapse, that something more definite may be heard.

In 1871, the scheme in which Sir Hugh Allan figured so prominently, but which, fortunately for the country, fell to the ground, involved a grant of 20,000 acres of land, together with a cash bonus of $12,000 for each mile of railway constructed. The land was to have been taken up along the entire length of the road from Nipissingue to the Pacific, good and bad acres, indiscriminately.

That scheme, impolitic as Canadians then judged it to be, was far less dangerous to the interests of the North-West than the present proposition. It now appears, if newspaper reports can be relied on, that English capitalists will not look at any of the lands within the Woodland and Rocky Mountain regions, knowing that both eastern and western sections of the road where located now, pass through a worthless country. They are to help themselves to the "cream" of the North-West, and will confine their choice within the erroneously designated "thousand mile" belt of prairie.

The proposition, if allowed, will be excessively unwise, and merits universal reprobation..

If the original scheme of 1871 offered speculators any real advantages—which from our present knowledge of the country, appears doubtful—the proposition of to-day, while trebling the apparent inducements to capitalists, so far as the lands are concerned, will be productive of the most disastrous effects throughout the North-West.

Of course, but for the harassing arrangement of 1871 with the Pacific Province, there would be no necessity or excuse whatever for such measures as are apparently in contemplation, because the Dominion Government, were it at liberty to carry on the work in a common-sense manner, is perfectly able to build the prairie sections of the Pacific Railway as fast as necessary, without overburdening the tax-payers of the older provinces, while the construction of the British Columbian portion of the road could be deferred; but politics, and the Pacific Province being paramount over all other considerations, the entire North-West may shortly be sacrificed on that account, and find itself bound hand and foot under the domination of a gigantic and soulless monopoly, unless the people awaken to a sense of the impending danger.

Let us enquire for a moment into the consequences of transferring the only available choice lands of the North-West from the custody of the Government to that of a great corporation or body of capitalists. It is universally admitted that all land monopolies are a curse, and utterly subversive of a fair and liberal policy. To-day in Manitoba, as the result of an atrocious system on a small scale, what between the Hudson Bay Company and some large private proprietors, it is impossible to purchase lands either at or near Winnipeg, or along the line of the Pacific Railway, excepting at ridiculously high figures. To such a degree has the abominable system been carried, that recently arrived intending settlers have turned back in disgust to take up the equally good, but cheaper lands of Dakota and Minnesota.

What then will be the state of things when the whole prairie belt is controlled by a private corporation? The inevitable result will be to unfairly enhance the price of all lands within the limits of the railway grant, and to impede or totally prevent settlement of the soil by the poorer classes of colonists to whom we must look in a great measure for agricultural development. In the case of the Central Pacific Railway, Congress granted to the Company all the alternate sections on each side of the road, for twenty miles back, or an area of 12,800 acres for each mile built. The immediate result was to increase the price of the ordinary public lands retained by Government within the limits of the railway grant, from $1.25 to $2.50 per acre. Congress was actually forced to raise the price of Government lands *at the bidding of the Railway Company*, which, of course, had no desire to see adjoining lands sold for half the price of the railway reserves. Similar, if not worse results will follow in the case of the Canadian Pacific Railway. The lands will be sold at a high figure, and in many instances on credit; they will be mortgaged to the Railway Corporation at high interest, and, the result in the majority of cases will be that the poor, struggling settler will insensibly drift into a state of bondage, while an odious system of feudalism will be inaugurated throughout the North-West.

Let any one enquire into the condition of the small farmers along the lines of railway in the United States, which have been subsidized by land grants, and it will be found that even in the best settled States of the west, a large number of the fairest farms are mortgaged beyond redemption to the grasping corporations which own and control the roads.

The unhappy effects of land monopolies must still be fresh in the memory of French Canadians. How were the townships between the St. Lawrence and the frontier depopulated ? What caused the exodus of the flower of Canadian youth, when one-twentieth of the whole population of Lower Canada, some thirty years ago, went into exile, driven from their homes by a selfish land policy, to increase the population of Maine and Vermont ? What was the testimony of the Abbé Ferland and many of the most respected Roman Catholic clergymen of Lower Canada, when questioned as to the causes of depopulation ? And will the French Canadians of to-day suffer the repetition of such a policy of extinction and degradation of the French race in the North-West, of which their daring ancestors and self-denying priesthood have been the honoured pioneers and discoverers ?

The Government is bound by every principle of justice to watch over and jealously guard Canada's heritage in the west. It is bound to see to the encouragement of the honest settlement of the country, and the only way to do this is to render the possession of extravagantly large estates burdensome to the owners, and to compel the occupation of the land by its proprietors. Any other land policy will bring about the most disastrous effects.

At present, owing to the extravagant ideas of which the public has become possessed through misleading reports with regard to the fertile areas west from Manitoba, it may be interesting to review briefly the character of the lands adjoining the proposed line of railway, from Nipissingue to the Pacific. The idea has gone abroad that the entire country from Red River to the Rocky Mountains is a perfect flower garden, inexhaustible as to its extent and resources.

People accustomed to the dark forests of Eastern Canada, are too apt to rush into ecstacies at the unwonted appearance of the grassy plains of Manitoba. In fact, there has been a little too much enthusiasm. What is the reality?

There are no good lands between Nipissingue and Red River, the region north of Lakes Huron and Superior, being for the most part rocky, sterile, and difficult for railway purposes, while the country between Thunder Bay and Manitoba, offers but few inducements for settlement.

The "Woodland Region," in the words of the Chief Engineer, "*does not offer any great prospect of becoming an agricultural country.*"

The maps facing pages 234 and 245 of the last Railway Report, are calculated to mislead, and the assertion is here reiterated, that the good saleable, agricultural lands, along the proposed railway line, cease at Lake St. Anne, some 45 miles west of Edmonton Fort. Between Lake St. Anne and the Yellow Head Pass—a distance of 200 miles—we have the authority of Engineer McLeod and of the Reverend George Grant, and others, for stating the country to be nearly all worthless and much of it muskeg, while, regarding the valueless character of the British Columbian section, the reader can refer to the authorities already quoted.

The only valuable lands along the route of the proposed Pacific Railway, are between Manitoba and Edmonton, or more strictly speaking, Lake St. Ann—a distance of about 750 miles—and the character of those lands has been thus described by Mr. H. MacLeod, an officer of Mr. Fleming's staff, and a gentleman of undoubted veracity.

" Of the country between Winnipeg and Lake St. Ann, I estimate
" that the proportion of excellent farming land is about 43 per cent.;
" fair land, 15 per cent.; and poor, light sandy, or clay and boulders,
" 42 per cent.; the latter is, however, suitable for grazing purposes.
" The hills are generally poor soil. The area of land covered with

" timber—small poplar—between Livingston and Edmonton, along the
" line of railway, is about 54 per cent. For 200 miles west of Living-
" stone the country is much covered with wood and water."

From the above quotation, it will be noted that, only about *one-half*
of those lands is really good, and readily saleable. The extracts from
other Surveyor's reports confirm Mr. MacLeod's statement.

In view, then, of the true facts of the case, in so far as the land
question is concerned, it may not be such an easy matter to induce
foreign capital to assume the responsibility of building the Pacific Railway, as now proposed. If the unwise scheme of conveying the choice
lands of the North-West to a foreign Company be carried out, and a
Corporation be found sufficiently insane to undertake the construction of
a railway through the Eastern Woodland region and the mountains of
British Columbia, in consideration of even such a large slice from the
central section, as the Government may dare to offer, and a cash bonus
such as that proposed in '1871, those who embark in the scheme must
bear the consequences. All the writer can say on the subject is " *Caveat
Emptor.*"

In the neighbouring states and territories of the Republic, there are
millions of acres of fine lands to be had for a merely nominal sum, and
companies of Capitalists would find, that to build the Pacific Railway
even from Manitoba to the Pacific, all the Dominion acres within 100
miles of the line of road in the fertile section, *i.e.* between the western
boundary of Manitoba and Edmonton, will not suffice. The proof is a
simple calculation. A strip of 700 miles in length, and 200 miles in
width, equals 140,000 square miles, or a little over 89,000,000 acres,
nearly half of which must be assumed of inferior quality, if we adopt Mr.
H. MacLeod's estimate of the proportions of inferior, to first-class fertile
areas. But the Government cannot give away such wholesale quantities
of land in this manner. The proposition is, it may be presumed, to grant
alternate sections only, so that the estimate now made must be reduced
one-half, and the probabilities are that, even upon the basis of " *hundred
mile* " blocks on both sides of the southern line, the choice lands available for capitalists will not exceed, after deductions for worthless lands
are made, an area of 25,000,000 of acres.

It has been shown upon the authority of Messrs. Dawson, Selwyn,
Marcus Smith, Macoun, Eberts, O'Keefe and Smith, that from the
" Middle Forks " of Pine River, upon the northern line, a nearly contin-

uous belt of fine agricultural and pastoral land stretches almost uninterruptedly to Manitoba, *a distance of one thousand miles.*

The estimates of Messrs. MacLeod and Cambie lead to the belief that the western section of this line, will cost much less than the corresponding section on the Yellow Head route. Mr. Marcus Smith has stated that it may be safely estimated that work of construction will generally cost much more on the southern than on the northern line. Mr. S. Fleming, the Chief Engineer, has given *his* opinion that the present work of construction, on the Frazer, is unwise, in view of the advantages offered by the northern line. The accumulation of evidence given in those pages confirms these views: The northern line is therefore the better of the two for purposes of colonization and also of construction. Canadians may note these facts.

The writer is perfectly aware that the views embodied in this pamphlet will create some surprise, and, perhaps, excite the indignation of those journals which have already constituted themselves the "Mentors" of the people upon the question discussed.

The writer has, as must be perfectly apparent, abstained as much as possible from obtruding his own views upon the public, his aim has been so to group all the trustworthy evidence as to afford those chiefly interested—the already overburdened taxpayers—ready means of making themselves thoroughly acquainted with the subject. It is claimed for this paper that it is more a synopsis of evidence, a summing up, as it were, of the best testimony, than an exposition of any particular theory. A simple, unvarnished statement of hard, stubborn facts has been made, and the writer frankly admits, that it will concern him much to see that statement unfairly impugned, as an attack upon this pamphlet can mean nothing more or less than an outrage upon many of the ablest and most valuable officers of the Geological and Pacific Railway Staff of Canada, whose official evidence has formed, to a great extent, the basis of this discussion.

www.ingramcontent.com/pod-product-compliance
Lightning Source LLC
Chambersburg PA
CBHW020337090426
42735CB00009B/1575